CHILTON'S Guide To:

EMISSION CONTROLS
and
HOW THEY WORK

ILLUSTRATED

Prepared by the

Automotive Editorial Department

Chilton Book Company

Chilton Way

Radnor, Pa. 19089

215—687-8200

editor-in-chief **JOHN D. KELLY;** managing editor **JOHN H. WEISE, S.A.E.;** assistant managing editor **PETER J. MEYER;** editor in charge **MILES SCHOFIELD;** technical editor **N. BANKS SPENCE, JR.;** editor **MARTIN W. KANE**

CHILTON BOOK COMPANY   RADNOR, PENNSYLVANIA

Published in Radnor, Pa., by Chilton Book Company
and simultaneously in Ontario, Canada,
by Thomas Nelson & Sons, Ltd.
Manufactured in the United States of America

Library of Congress Cataloging in Publication Data

Chilton Book Company. Automotive Editorial Dept.

Chilton's Guide to Emission Control and How They Work.

L.C. 74-6332
ISBN 0-8019-6083-5
ISBN 0-8019-6084-3 (pbk.)

ACKNOWLEDGEMENTS

Many individuals and companies have contributed data and illustrations for this book. Whenever possible, the names of companies were listed near their data or illustration. Our thanks to those who have contributed, especially those individuals who worked hard to answer technical questions. It is only through the help of many dedicated people in the auto industry that compilation of such a book is possible.

Although information in this guide is based on industry sources and is as complete as possible at the time of publication, the possibility exists that the manufacturer made later changes which could not be included here. While striving for total accuracy, Chilton Book Company can not assume responsibility for any errors, changes or omissions that may occur in the compilation of this data.

Contents

Chapter One

Introduction

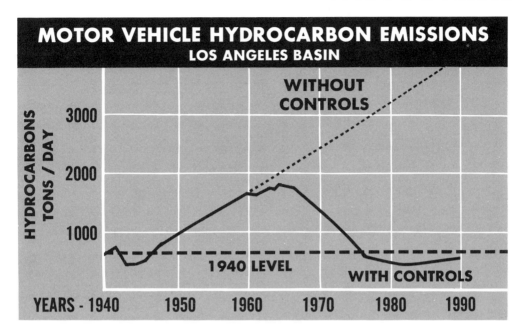

MOTOR VEHICLE HYDROCARBON EMISSIONS
LOS ANGELES BASIN

The Origin of Smog

It wasn't too many years ago that smog was considered a local problem in the Los Angeles area. Now we all know differently. There is smog almost any place where there is a large concentration of motor vehicles, and industry.

The internal combustion engine puts three kinds of pollutants into the air, two of them are responsible for smog. The three pollutants are hydrocarbons (HC), oxides of nitrogen (NOx), and carbon monoxide (CO).

HYDROCARBONS

Hydrocarbons is really just a fancy word that means unburned gasoline. If you take a can of gasoline, pour it out on the ground, and let it evaporate, the vapors that go off into the atmosphere are hydrocarbons. Because hydrocarbons are created by the simple evaporation of gasoline, a motor vehicle can pollute the air even when it's not running. This is the reason for the evaporative vapor controls which trap the vapors so they cannot escape into the atmosphere.

If hydrocarbons can be burned they don't cause any problem. Light a match to that gasoline that you poured out on the ground and allow it to burn up completely. The result will be nothing but carbon dioxide and water, which is harmless. The hydrocarbons have been eliminated by burning.

If the internal combustion engine would burn gasoline completely, all the hydrocarbons would be gone and we would get carbon dioxide and water out of the tail pipe. The sad story is that nobody has yet been able to make an internal combustion engine that will give perfect combustion.

It is hard to believe that you can put gasoline into a combustion chamber, burn it with enough heat to drive an automobile and still have unburned gasoline coming out of the tail pipe. Actually the combustion takes place in the combustion chamber so fast, that there is a layer of gasoline around the edges of the chamber that doesn't get burned. This happens not only in the combustion chamber of a car, but in any burning within an enclosed space. The big guns of a battle-

1

ship are loaded with a projectile backed up by bags of powder. After the gun is fired you can actually crawl into the firing chamber and find pieces of powder and even parts of the cotton bags that did not burn up. On a much smaller scale this happens even in a gun cartridge although you may not see the unburned powder because most of it goes out the barrel with the bullet.

CARBON MONOXIDE

Hydrocarbons that burn up in the combustion chamber are eliminated as far as pollution is concerned, but the way in which they burn determines how much we get of another pollutant, carbon monoxide. A rich mixture burns with too much gasoline, or if you want to look at it the other way around, you could say there is not enough air. When burning takes place the hydrocarbons unite with oxygen. If we have a rich mixture, there is not enough oxygen and so we get carbon monoxide (CO). Carbon monoxide is a deadly poison, the stuff that kills you if you run your car in a closed garage.

The formation of carbon monoxide in the combustion chamber can be cut down by running lean mixtures. This way the combustion is more complete and the end result is carbon dioxide (CO_2), which is harmless. Lean mixtures have been used by the car makers for several years to cut down on carbon monoxide emissions.

OXIDES OF NITROGEN

The third emission, oxides of nitrogen is created when the peak temperature in the combustion chamber rises over 2,500 degrees Fahrenheit.

We ordinarily think of the air we breathe as being mostly oxygen, because that is the gas we need to stay alive, but actually the air around us is 80% nitrogen. With that much nitrogen in the air that the engine breathes, naturally there is plenty of it in the combustion chamber when the spark plug lights the fire. If the peak temperature in the combustion chamber is 2,500 degrees or hotter, the nitrogen and oxygen in the air combine to form oxides of nitrogen. NOx also forms at lower combustion temperatures but in smaller amounts.

PHOTOCHEMICAL SMOG

It is important to remember that each of the three pollutants is emitted for separate reasons. Hydrocarbons come from unburned fuel, carbon monoxide comes from rich mixtures, and oxides of nitrogen come from too much heat in the combustion chamber.

These three pollutants are not smog. Smog is formed in the atmosphere in a photochemical process by only two of the pollutants after they leave the tail pipe. When hydrocarbons and oxides of nitrogen are exposed to sunlight, chemical reaction creates photochemical smog, which is a whole series of chemicals, called oxidants. It is these oxidants that create all the discomfort and even death when people or plants breathe them.

Carbon monoxide does not enter into the production of smog. It is a deadly gas all by itself, but it is completely odorless so you don't even know it is around until its too late.

THE SMOG IN LOS ANGELES

Los Angeles is blessed with probably the most livable climate in the world, with sunshine possible any day of the year, and temperatures that hardly ever get below freezing.

Unfortunately, nature created a few flaws in the paradise, mainly a range of mountains to the north that create a sheltered area known as the Los Angeles basin. The lid on the basin is the temperature inversion which is present in the Southern California area over three hundred days out of every year.

TEMPERATURE INVERSION

In most parts of the world, the air near the ground is warm and it gets cooler the greater the altitude. The

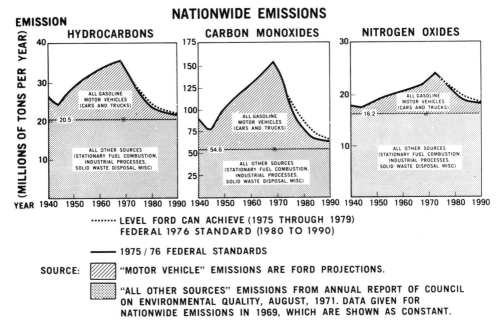

The nationwide emission picture is improving. Ford Motor Co. prepared this chart to show what they could actually achieve, compared to the original 1975-76 federal standards

warm air rises, meets cooler air and cools itself off but the air higher up is still cooler, so the air keeps rising until pollution is carried away. A temperature inversion is a warm air blanket extending from street level up to a few hundred, or even a few thousand feet. Because there is no cooler air at higher levels within the inversion, the air in the inversion becomes a stagnant mass that does not rise. When the top of the warm air blanket is several thousand feet high, there is enough movement within the inversion that the smog rises above the city streets and you have a clear day. When the inversion level is down to where it can be measured in hundreds of feet, the smog builds up and can't go anywhere. This is when you get the eye irritation, headaches and difficult breathing that is characteristic of a smog attack.

Now that the scientists have done all the work and positively identified where the smog comes from, it is very easy to explain it, but you should remember that it took many years of laboratory experiments to pin down the internal combustion engine as being responsible for the production of photochemical smog. Even

today, laboratory scientists do not know exactly how photochemical smog is formed or what it consists of. They can put the proper ingredients into a chamber, expose it to sunlight and create smog, but nobody has yet discovered the precise chemical reaction that actually goes on. Experiments are continuing and some day we may be able to get rid of smog simply by spraying something in the air. Until that time comes, the only way to stop it is to keep the automobile from emitting the pollutants that create it.

THE DEVELOPMENT OF CRANK-CASE, EXHAUST, AND VAPOR CONTROLS

When work first started in smog research, most people naturally assumed that all the pollutants came out of the tail pipe. The car makers themselves did some experimenting and pointed out that the crankcase road draft tube was responsible for 20% of the hydrocarbon emissions.

Crankcase Controls

In an engine, some of the hydrocarbons that cling to the combustion chamber walls and do not burn, are forced out through the exhaust valve

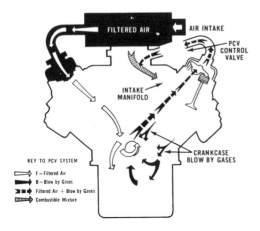

KEY TO PCV SYSTEM

⟹ F = Filtered Air
➡ B = Blow by Gases
➤➡➡ Filtered Air + Blow by Gases
⟼ Combustible Mixture

The closed system results in 100% control of crankcase emissions

when it opens. But there is also a layer of hydrocarbons around the top ring in the space between the top of the piston and the cylinder wall. There never has been a top ring that made a perfect seal, so there is always some leakage or blowby past the ring and down into the crankcase. Because the space above the ring is full of unburned hydrocarbons, these are what go down through the crankcase and out through the road draft tube. There is some NOx and CO mixed in, but crankcase emissions are mostly HC.

Starting on 1961 California cars, and in 1963 nationwide, the manufacturers removed the road draft tube from the crankcase and connected the crankcase opening to the intake manifold through an orifice or a movable valve to control the flow. This positive crankcase ventilation system, or PCV, completely eliminated the emission of hydrocarbons from the crankcase, except at wide open throttle where there is no intake manifold vacuum to pull the pollutants into the engine. At wide open throttle the engine breathed through an open filler cap on the oil filler tube.

In 1966, as a result of legislation, the manufacturers put a closed PCV system on their California cars and in 1968 this closed system was installed nationwide. With the closed system there is absolutely no escape of hydrocarbons into the atmosphere.

At wide open throttle the crankcase emissions are drawn into the engine by the suction in the air cleaner.

Exhaust Controls

Starting in 1966, on California cars, the car makers attacked the problem of exhaust emissions, concentrating mainly on hydrocarbons and carbon monoxide. The easiest way to control carbon monoxide is to run lean mixtures. Most carburetors since 1966 in California or 1968 nationwide have been set on the lean side. Hydrocarbons were controlled partly by changing the shape of the combustion chamber so there were fewer pockets where the gasoline could escape burning.

Unfortunately, those engine modifications did not do a complete job for most of the car makers, and many of them, in the early days of emission controls, had to use an air pump, which takes care of both hydrocarbons and carbon monoxide. The pump, driven by a belt off the front pulley, pumps air through a series of hoses and lines into each exhaust port. Any hydrocarbons or carbon monoxide that come out of the port are oxidized (burned) when they mix with the blast of air under the heat of the exhaust. Actually, the system creates combustion in the exhaust manifold and if you don't believe it, just get under a car equipped with an air-pump after a hard run at night and watch how everything glows.

Chrysler Corporation avoided the use of an air pump for many years, by using an engine modification system. Later on, other car makers also adopted the engine modification system. They were able to do this by very carefully tailoring the spark timing and the mixture so that fewer hydrocarbons and carbon monoxide were produced. What did come out the exhaust port was burned up in the exhaust manifold because of retarded timing which heated up the exhaust.

Heat in the exhaust manifold or exhaust system gets rid of carbon monoxide and hydrocarbons by burning them up, but does not produce oxides of nitrogen because it does not get hot enough.

The crankcase controls that began in 1961 and the exhaust controls that started in 1966 were mainly concerned with hydrocarbons and carbon monoxide. Oxides of nitrogen were recognized as a pollutant that created smog, but hydrocarbons and carbon monoxide were much more easily controlled with the knowledge that was available back then. Public health officials legislated against the emission of hydrocarbons and carbon monoxide, completely ignoring the oxides of nitrogen. This blunder was compounded when the hydrocarbon and carbon monoxide controls came in because the lean mixtures raised peak combustion temperatures to above 2,500 degrees causing emissions of more oxides of nitrogen. Los Angeles residents found out about this the hard way when oxides of nitrogen suddenly became a much greater problem than they ever had been and the familiar brown haze that came with some smog attacks started appearing everywhere. Finally, in 1971, legislation forced the car makers to put NOx controls on new cars. That was the beginning of the transmission controlled spark and speed controlled spark systems, which lowered peak combustion temperatures by retarding the spark. General Motors put the systems on most of their cars in 1970, one year before the law actually required them. Lately, exhaust gas recirculation (EGR) has taken the place of transmission controlled spark. EGR lowers the peak combustion temperature because the exhaust gas will not burn, and it makes the fuel air mixture less powerful.

In the meantime, Californians were being smothered by the exhaust from those 1966 through 1970 vehicles which put out much more oxides of nitrogen than any of the older uncontrolled cars did.

California now requires that devices be fitted to those 1966 through 1970 used cars that do not have oxides of nitrogen controls. California also requires that some type of used car emission control be installed on

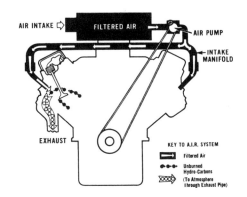

The air pump causes combustion in the exhaust system that burns up the emissions, making them harmless. Every air pump system also includes some engine modifications

vehicles as far back as 1955. So far, California is the only State to require this but undoubtedly other States will, if their smog problem gets bad enough.

Vapor Controls

Starting on 1970 California cars, the manufacturers installed vapor controls to stop the evaporation of gasoline from the fuel tank and the carburetor. The first two years, some systems used the engine crankcase to store the vapors that came out of the fuel tank. When the engine was running, the vapors were purged from the crankcase by the positive crankcase ventilation system. This crankcase storage could cause trouble on cars that were driven on short trips because they never got warm enough to evaporate the gasoline in the crankcase. The PCV system was constantly sucking in fuel vapors that made the engine run rich.

In later years, most car makers, except some imports, went to a canister storage system. A canister in the engine compartment collected the vapors and was purged when the engine was running. The vapors from the canister didn't cause trouble in the operation of the engine because they were drawn off very slowly.

CONTROLS VERSUS DRIVEABILITY

The emission controls on the 1971 through current vehicles are doing a good job of controlling emissions, but the car doesn't run as well as it used to. Drivability has suffered. Surging and hesitation are expected. In fact, if you drive a late model car that does not surge or hesitate at *any* speed, then someone has probably doctored the car to make it run better.

Gas milaege has not been affected as much by emission controls as you might think. Other factors such as the increased weight of vehicles and the number of power accessories, such as air conditioning, take much more out of gas mileage than the emission controls do.

SYSTEMS OF THE FUTURE

The cars that we have seen since 1971 are just about what we will continue to see as long as the manufacturers keep using standard carburetion and ignition. Some manufacturers have already taken a step in the right direction by going to electronic ignition. But there will have to be better ways developed of feeding the fuel into the engine so that it only gets enough for perfect combustion and not enough to pollute. Until that happens, perhaps with some type of fuel injection, or a different type of carburetor, we will not have a car that drives well and is clean, at the same time, unless the catalytic converter saves us.

Catalytic Converters

The catalytic converter looks like a muffler full of beads. The beads are the catalyst, a substance that causes a chemical reaction, but does not enter into it.

Usually, the catalytic converter is hooked up to an air pump, similar to the pump on air injection systems. When the hot exhaust gases mix with the air in the presence of the catalyst, the HC, CO and NOx are reduced to harmless gases.

The catalyst eventually gets dirty or contaminated by foreign substances from the fuel tank. The fuel has to be absolutely lead free, or the catalyst will be ruined. If it is necessary to change the catalyst, the plan is to have a trapdoor in the converter, and a special vacuum cleaner. Connect the vacuum cleaner to the converter and it sucks out the old catalyst. The new catalyst is inside the vacuum cleaner. By switching the connections, the new catalyst is injected into position inside the converter. It only takes a few seconds with the car on a hoist.

Hopefully, cars equipped with catalytic converters can run rich mixtures for better driveability, and the catalyst will take care of the pollution. We can only wait and see.

Thermal Reactor

The thermal reactor goes on the car's engine in place of the regular exhaust manifold. It is used to oxidize unburned hydrocarbons and carbon monoxide before they can be released into the atmosphere.

Basically, the thermal reactor is an extension of the air injection system concept. What a thermal reactor does is supply a better place for the exhaust combustion to take place (Remember the glowing exhaust manifold above?) Combustion is aided because there is more time, turbulence, and higher temperatures to burn up the unburned hydrocarbons and carbon monoxide in the exhaust.

Exhaust gases, mixed with air from an air injection pump, are fed from the exhaust ports into the reactor. Baffles at the end of the reactor force the gases back to the center, thus giving them more time in the high temperature area and added turbulence to provide for their more complete combustion. The reactor is insulated, both to keep its internal temperature high and to keep things from getting too hot under the hood.

Stratified Charge Engines

A stratified charge gasoline engine works something like a diesel engine. Two types of mixture are provided in the combustion chamber: a small, very rich cloud of mixture to promote ignition and a larger, very clean cloud of mixture which contains an excess of oxygen, to aid in more complete combustion.

There are at least two different ways of providing the two different kinds of mixture. Either a fuel injection system having two nozzles, one for rich mixture and one for lean, per cylinder or a separate, small combustion is used for each cylinder, in which the rich mixture is ignited to help more completely burn the leaner mixture in the main combustion chamber by providing sufficient heat to raise all portions of the chamber to above ignition temperature. In this way, a mixture too lean to burn in an ordinary engine may be burned, resulting in lower exhaust emissions.

Fuel Injection

Fuel injection may also be used to help decrease the amount of emissions, by providing a more precise control over the fuel/air ratio for each cylinder. In a normally carbureted engine, the mixture varies between cylinders because of their varying distances from the carburetor. As a result, some cylinders get a very rich mixture, while others may get a very lean mixture. So, at best, the carburetor mixture setting is a compromise.

When the fuel in injected directly into the combustion chambers, it is possible to provide identical (and ideal) mixtures to each cylinder.

Also, fuel injection provides a higher fuel delivery pressure which causes better atomization (smaller droplets) of fuel and thus more complete combustion.

With the more complex electronic fuel injection systems, it is possible to tailor the mixture for such variables as engine load, ambient temperature, etc., allowing an even better control of the mixture.

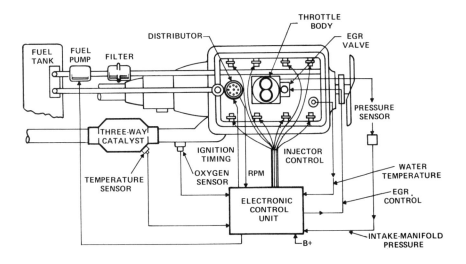

A closed-loop EFI system for 1976 emission standards. In order for the catalytic converter to operate effectively, the exhaust chemistry must be strictly controlled. This requires an exact metering of the air/fuel intake ratio through all phases of engine operation, which in turn requires a constant sensing of various engine conditions and exhaust characteristics. This is all fed to the Electronic Control Unit which computes the engine's exact fuel requirements.

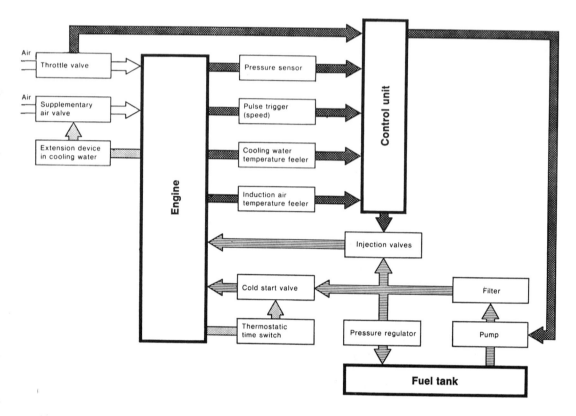

Mercedes-Benz electrically controlled injection system (Courtesy Mercedes-Benz of North America Inc. and Daimler-Benz AG. Stuttgart, Germany)

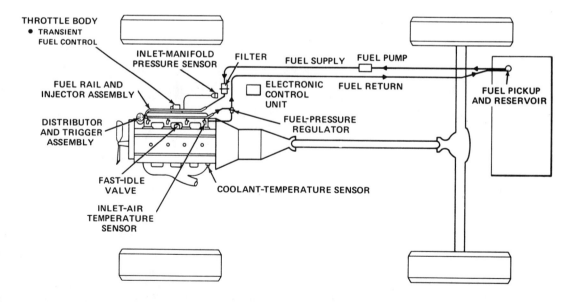

A typical Electronic Fuel Injection (EFI) system installation. The various sensors relay information on applicable engine operating conditions to the Electronic Control Unit. The control unit in turn activates the fuel pump, insuring that the necessary amount of fuel is delivered to the fuel injectors. Any excess fuel is returned to the gas tank via an overflow valve and fuel return line.

Chapter Two

Emission Controls, How They Work

The "muffler full of beads" may be the solution the car makers have been looking for. Its appearance on 1975 cars may mean better driveability

Original Equipment Systems

CRANKCASE VENTILATION CONTROLS

Road Draft System

Because piston rings don't seal perfectly there is always a little bit of leakage or "blow-by" into the crankcase. This blow-by consists of unburned gasoline and water vapor, which must be removed or the oil pan ends up with corrosive acids from chemical reactions between the water and the gasoline. Engineers recognized a long time ago that the oil and the engine would last longer if they kept the crankcase ventilated. The way it was always done before smog was with an opening somewhere at the top of the engine and a tube off the side of the crankcase that hung down into the airflow under the car. The airflow, or road draft, passed over the tube when the car was moving, drew the undesirable fumes out of the crankcase and pulled in fresh air through the opening on the top. Road draft tubes worked well, while the vehicle was moving, but on a stationary engine or one that didn't move very fast, the engineers had to devise other ways such as fans or vacuum systems to keep the crankcase ventilated. In the days before

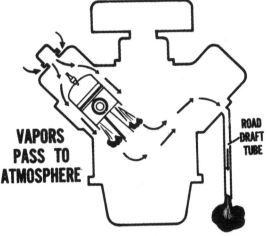

VAPORS PASS TO ATMOSPHERE

ROAD DRAFT TUBE

The old road draft system of crankcase ventilation worked well as long as the car was moving, but at idle it didn't work at all

smog, it didn't matter how you did it just so you had a blast of clean air going through the crankcase to get rid of the fumes.

Open Positive Crankcase Ventilation (PCV)

Positive crankcase ventilation uses intake manifold vacuum to draw the fumes out of the crankcase and into the manifold. They are burned in the combustion chamber and go out with the engine exhaust. The crankcase connection can be any place on the engine that has internal air passages connecting with the crankcase. The original systems were an add-on de-

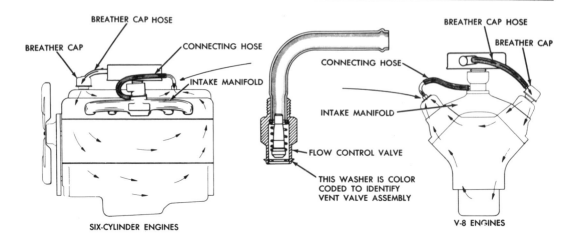

The first positive crankcase ventilation systems took in fresh air through an open breather cap

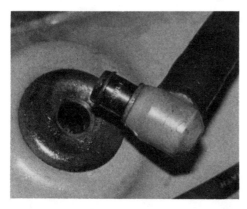

The PCV valve is usually plugged into a grommet on top of the engine. This is a late model Chrysler valve, which looks much different from the earlier hooded models

vice connected at the same position as the old road draft tube. The hose ran from there up to the intake manifold. Later on, when the systems were designed into the engines, they found that they could connect to the crankcase at the rocker cover, at the valley cover, at either end of the block below the intake manifold or even at the fuel pump mounting flange. The crankcase connection should be where there is the least chance of picking up any oil, so it is usually made at the highest point, which is the rocker cover.

In 1971, Ford moved all of their crankcase connections from the rear of one rocker cover to the front of the opposite cover. This didn't affect the working of the system, but they found that they picked up less oil when the connection was at the front.

For years, Pontiac has had their crankcase connection at the front of the valley cover, underneath the intake manifold.

From the crankcase connection most PCV systems use a hose or tube that runs to the intake manifold. But you can't just connect a big hose and let it go at that. There has to be some kind of restriction so that the engine doesn't pull more air out of the crankcase than it does through the carburetor. This restriction can be a plain orifice or a variable orifice, which is called the PCV valve. During the high vacuum at idle or deceleration, the plunger in the valve is pulled against a spring and seats against the end of the valve. In this position the amount of airflow through the valve is limited. At cruising speed, when intake manifold vacuum is a little bit less than it would be at idle, the spring pushes the valve off its seat, making the opening slightly bigger and allowing more airflow. At wide-open throttle, the spring pushes the valve completely off its seat, but because there is no vacuum at wide-open throttle there is no airflow through the valve at that time.

All PCV valves will close if there is an intake manifold explosion or intake backfire. When the intake manifold pops back through the carburetor, the pressure in the manifold increases and pushes the PCV valve plunger to the end, preventing this pressure from going down into the crankcase. The fixed orifice provides a similar amount of protection because the intake manifold explosion cannot get through the orifice fast enough to build up pressure in the crankcase. What is feared is that flame might travel along the hose and cause a crankcase explosion. This is impossible as long as the orifice or the PCV valve is in place.

Part of the positive crankcase ventilation system is the fresh air entry. There usually is an opening somewhere on the engine to allow fresh air to enter. Some car makers, particularly those who make import cars with small engines, have PCV systems without a fresh air entry. In that type of system the crankcase runs under vacuum at all times and there is very little flow through the PCV connection. They can get away with this because small engines do not have much blow-by. At wide-open throttle when there is no intake manifold vacuum to apply to the crankcase, the vapors are pushed into the intake manifold by crankcase pressure. This system obviously does not keep the crankcase purged of fumes as well as the type that has the fresh air entry.

In the open system the fresh air entry is usually the oil filler cap, although it can be a separate breather on the rocker cover. In that case, the oil filler cap is the solid type. Inside the fresh air breather cap or separate breather is a mesh type filter that keeps dirt from entering the crank case.

During wide-open throttle on the open system, there is no vacuum and the crankcase breathes from its own buildup of pressure, through the oil filler cap or breather. This system is not 100% perfect because at wide-open throttle it does allow crankcase vapors to escape into the atmosphere.

Closed Positive Crankcase Ventilation (CPCV)

If you want a completely closed system with no leakage of crankcase fumes to the atmosphere, then you have to use what is known as closed positive crankcase ventilation. The PCV valve and the hose from the crankcase to the intake manifold are the same on both open and closed systems. The only difference in the closed system is the way that the fresh air entry is hooked up. Fresh air must enter from the air cleaner into a hose or other connection to the crankcase. Usually the system is set up so that fresh air enters a PCV filter inside the air cleaner, then goes through a hose to the rocker cover.

The advantage of a closed system is that it does not pass fumes from the crankcase into the atmosphere at wide-open throttle or if the PCV valve gets clogged. When the engine is running with intake manifold vacuum, the airflow through the fresh air hose is from the air cleaner to the rocker cover. When the engine is running without manifold vacuum such

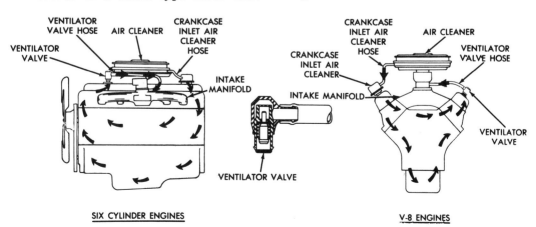

SIX CYLINDER ENGINES   V-8 ENGINES

Later systems were closed, with fresh air coming from a tube that connected to the air cleaner

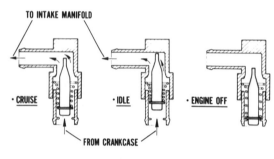

PCV VALVE

The jiggler pin inside the valve takes a position according to manifold vacuum. During an intake manifold explosion (popping back through the carburetor) the valve goes to the "engine off" position and prevents the burning mixture in the manifold from traveling down the hose into the crankcase

as at wide-open throttle or if the PCV valve should be plugged, then airflow is by crankcase pressure through the fresh air hose from the rocker cover back to the air cleaner. Once the fumes get inside the air cleaner they are sucked down into the engine, so that there is no way that the crankcase fumes can get outside the engine and into the atmosphere.

California requires that some older cars have their open type PCV systems converted to a closed type. It is not necessary to buy a complete PCV system to do this because the manufacturers make a kit consisting of a fresh air hose and connections between the air cleaner and the rocker cover.

Some imported cars use a closed PCV system that does not have a valve. There is simply a hose connecting the crankcase to the air cleaner. A slight amount of suction in the air cleaner pulls the vapors out of the crankcase, but most of the flow is from crankcase pressure pushing out through the hose. In this system, as in all closed systems, the oil filler cap is the solid type.

Servicing Crankcase Ventilation Systems

Service on crankcase ventilation systems amounts mainly to cleaning the hoses, replacing the valve, and either cleaning or replacing the PCV filter. The car makers do not recommend that you attempt to clean the PCV valve because you cannot get it

apart to find out if your cleaning was successful. Some of the older cars had valves that could be taken apart and it is alright to clean those types because you can see whether you've gotten them clean.

PCV systems give very little trouble, but there is one thing you must watch if you are replacing a PCV valve with a valve that is not original equipment. Some of the valves on the market are universal, made to fit just about any engine and can easily be installed backwards. A valve that is installed backwards will be closed all the time because intake manifold vacuum will suck the plunger against the end of the valve away from the spring.

If you use a non-standard PCV

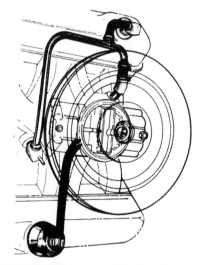

Earlier PCV systems used tubing to take vacuum from each side of the manifold to eliminate pulsing at the valve. Later systems made the connection underneath the carburetor, which accomplished the same purpose without all the tubing

valve, make sure that it is specified for your particular engine. The amount of air flowing through the valve is determined by the size of the plunger and the holes inside. You can't get inside to measure it, so that you have to rely on the part number.

If you have an older car with so much blow-by that the original equipment PCV valve can't handle the smoke, a universal PCV valve may be the answer. Some of the universal

This accessory PCV valve is not original equipment, but it has a high flow which can solve the problem of excessive smoke from the crankcase

The PCV filter is usually at the end of the fresh air hose, inside the air cleaner

valves pass so much air that you can actually run the engine on what comes up out of the crankcase with the carburetor mixture screws screwed all the way shut.

On an open system, a PCV valve that cannot handle the blow-by will show up in smoke and fumes coming out through the oil filler cap. On a closed system, you probably won't be able to see any fumes coming out of the air cleaner, but the air filter element will get dirty in a very short time because it is being forced to gulp oil fumes from the crankcase.

For many years Chrysler used a foam wrapper around their paper air cleaner elements to keep the oil off the element. A similar wrapper is now being used on other makes.

AIR PUMP EXHAUST CONTROLS

Air pump systems have a lot of plumbing, with hoses and lines run-ning all over the engine which make them look very complicated, but actually, they are one of the simplest systems for emission control. The pump, driven by a belt at the front of the engine, pumps air under a pressure of only a few pounds into each exhaust port. The hydrocarbons and carbon monoxide that come out the port are very hot. The extra air mixed with them causes a fire in the exhaust manifold that oxidizes the carbon monoxide into harmless carbon dioxide and burns up the hydrocarbons into carbon dioxide and water. Stainless steel nozzles are used to direct the air into the port as close to the exhaust valve as possible. The stainless steel is necessary so that the nozzles will not burn away.

Between the nozzles and the pump is a check valve. The system can be set up so that there is one check valve per bank on a V8 engine or a single check valve for the whole system. The check valve keeps the hot exhaust gases from flowing back into the pump and hoses, and destroying them. The pump parts are made out of a special hard plastic. If you have ever smelled a pump that burned up because of a bad check valve, you'll never forget it.

During closed throttle deceleration, high intake manifold vacuum pulls a lot of extra fuel into the engine and out the exhaust system. If the pump continues pumping during decelera-tion you can get a lot of popping in the exhaust or even an exhaust explo-sion that can blow the mufflers apart. To prevent this, some way had to be found to shut off the pump during de-celeration. The early systems used on most cars in 1966 and 1967 (and on some imported cars even later), used what was known as an anti-backfire valve or gulp valve. The gulp valve was connected between the pump and the intake manifold. A small sensing line or hose led from the valve dia-phragm to the intake manifold. Dur-ing the high vacuum of deceleration, the vacuum through the sensing line acted on the diaphragm, which pulled the valve open, allowing all of the air from the pump to flow directly into

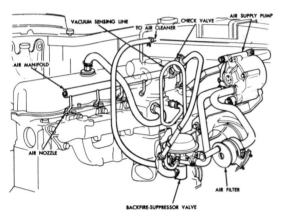

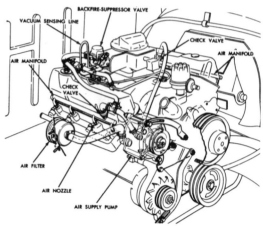

Early air pump systems had complex hose routings. Even so, they gave very little trouble in service

The backfire supressor valve and separate pump air filter have both been done away with on the later systems, except on some imported cars

the intake manifold. The anti-backfire valve did not shut off the flow of air to the nozzles, but because it opened the system to the high intake manifold vacuum, almost all of the air from the pump flowed into the intake manifold with very little left to come out of the nozzles. Limiting the air going into the exhaust manifold prevented exhaust system explosions and the extra air entering the intake manifold leaned down the mixtures so that emissions were not so bad during deceleration. The disadvantage of the anti-backfire valve was that the engine kept on running during deceleration. In some cases backing off the throttle at 70 miles per hour would have no effect at all for a few seconds. It felt as if the throttle was stuck wide-open. This was quite a shock if you weren't expecting it, but

even after you got used to it, it was still annoying to have a car feel like it was running away with you. Another disadvantage of the gulp valve was its tendency to open when the car first started. The gulp valves operate on a change in vacuum applied to the diaphragm. The change from no vacuum when the engine is not running to immediate vacuum when the engine starts was enough to open the valve for a few seconds. Many engines would shake because of the air entering the intake manifold immediately after they were started. Oldsmobile recognized the problem on V8 engines and provided an air bleed control valve that blocked off the air from the pump to the gulp valve until it built up to a high value. The gulp

The vane is travelling from a small area into a larger area—consequently a vacuum is formed that draws fresh air into the pump.

As the vane continues to rotate, the other vane has rotated past the inlet opening. Now the air that has just been drawn in is entrapped between the vanes. This entrapped air is then transferred into a smaller area and thus compressed.

As the vane continues to rotate it passes the outlet cavity in the pump housing bore and exhausts the compressed air into the remainder of the system.

Operation of the eccentric vane air pump. The spring-loaded gadget at the upper right is the relief valve, which only opens under high pressure

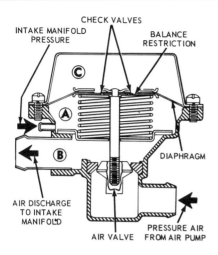

Note the "balance restriction" hole in the diaphragm of this backfire supressor valve. That hole keeps the valve from opening unless there is a sudden increase in manifold vacuum. Later diverter valves work on the same principle

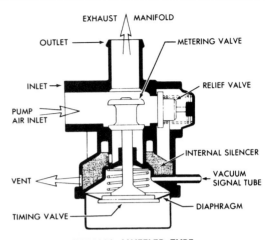

INTERNAL MUFFLER TYPE

Rochester Products latest diverter valve has an internal silencer. The bleed hole in the diaphragm does not show here, but it is definitely there

valve still opened when the engine started, but no air could enter the intake manifold because the control valve was closed.

The gulp valve was obviously not the way to go, so the next development became known as the dump valve, by-pass valve, or diverter valve. It used a diaphragm operated from intake manifold vacuum, the same as the older gulp valve. However, air from the pump ran through the diverter valve at all times. During the high vacuum of deceleration, the diverter valve shut off the air to the nozzles and sent it to the air cleaner. The air entering the air cleaner has no effect on the operation of the car. It was sent to the air cleaner just to cut down on the noise. Later diverter valves, and the ones that are still used today, exhaust the air directly into the atmosphere through either a bronze or a cotton filter.

In both the gulp and the dump type of backfire suppressor valves, the diaphragm has a calibrated hole in it. This means that if you apply a sudden amount of vacuum to the sensing nozzle on the valve, the diaphragm will move against the spring to operate the valve. The hole in the diaphragm bleeds off the vacuum very quickly and the spring pushes the

diaphragm back to the off position. Because of this bleed feature in the diaphragm, you can check the operation of the valve by disconnecting and reconnecting the hose or by simply pinching the hose and then letting it go. When you pinch the hose, you have to wait a few seconds for the vacuum to stabilize on each side of the diaphragm. When you reconnect the hose, the sudden surge of vacuum will make the valve operate. Of course, you can also check the valve by opening the throttle and allowing the engine to decelerate. In most cases, if you can reach the valve, or listen to the valve and work the throttle at the same time, making the engine decelerate is the easiest way to check the valve because you don't have any hoses or lines to disconnect.

When the pumps first came out, a relief valve was pressed into the side of the pump. The relief valve would open under high rpm to prevent the buildup of excess pressure that might damage the hoses or the pump itself. On some later models, the relief valve is built into the diverter valve, so that the diverter valve actually has a dual function. It diverts and also relieves pressure. If anything goes wrong with the relief valve part of the diverter valve, the entire diverter valve has to be replaced.

All domestic cars use a pump that is manufactured by the Saginaw Division of General Motors. The only

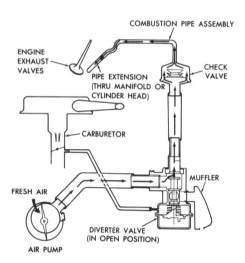

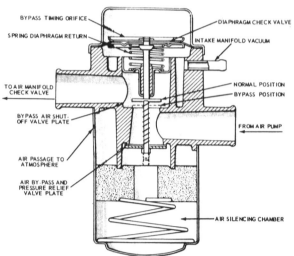

Late model air pump systems are much simplified from the earlier tangles of tubing and hose. This basic drawing applies to all domestic cars

Carter Carburetor Co. makes this diverter valve, which is used on Ford products. A little known feature of this valve is that it not only diverts, but will relieve high pressure as well

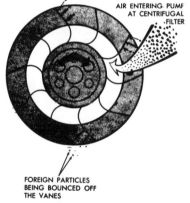

Diverter valves can be checked with the engine idling by pinching the hose, waiting a few seconds, and then releasing it. The valve should exhaust air for a few seconds

Operation of the filter fan on late model air pumps

variation in the pumps is the location of the mounting ears and the air hose connections. The early pumps used in 1966 and 1967, and on some cars for a few years after that, were made with 3 vanes and rebuilding of them was encouraged. Saginaw furnished parts for the 3-vane pumps until 1970 and then the parts were discontinued. Many mechanics had a lot of trouble reassembling the pumps because of the eccentric arrangement of the vane. In most cases the pumps were being replaced anyway, so that the loss of the supply of parts affected very few shops.

The 3-vane pump was discontinued on most new cars in 1968 when Saginaw came out with a new 2-vane design. The 2-vane pump has never had

any parts supply and it is factory-recommended procedure to replace the pump rather than attempt to repair it.

The 3-vane pumps took their fresh air supply from a separate air filter or from the clean side of the engine air cleaner. The most noticeable feature of the 2-vane pump was the elimination of any external air cleaner. The filter fan behind the front pulley of the pump acted as a centrifugal filter and kept dirt from getting into the inside of the pump. Cleaning of the filter fan is not recommended because it is too easy to get the dirt particles down inside of the pump. If the fan is so dirty that the air flow is restricted, then the

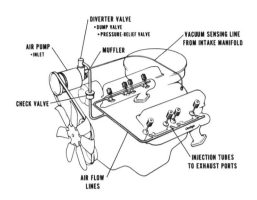

Chrysler 1972 air pump location

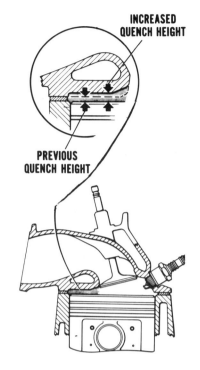

Increased quench height helps emissions by eliminating the "pocket" where unburned gasoline collects. Increased quench height has some side effects, such as requiring lower compression ratios, or higher octane gas

pulley and filter fan should be removed and a new filter fan pressed on the hub.

It is a difficult job to get the filter fan off the hub without breaking it. Actually, it is designed so that you do have to break the fan to remove it. Anything that attaches to the outside of the 2-vane pump, such as a relief valve, a diverter valve, mounting brackets, the filter fan, the front pulley or any air nozzle connections, can be replaced if necessary and they are available if you order them from a dealer parts department. The internal parts of the 2-vane pump are not available and it should not be taken apart.

The early pump systems used a lot of hoses and tubing, so many that in some cases that you couldn't even see the engine. Some car makers have stuck with the external lines and hoses, but others have built the air passages into the exhaust manifold, the intake manifold, or the cylinder head.

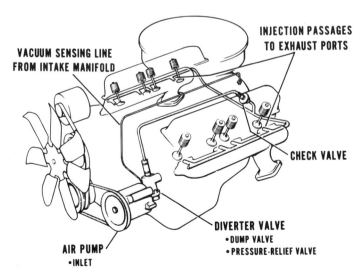

Chrysler 1973 air pump location

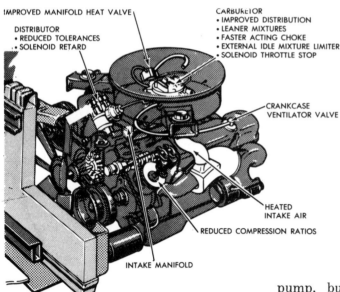

IMPROVED MANIFOLD HEAT VALVE

DISTRIBUTOR
• REDUCED TOLERANCES
• SOLENOID RETARD

CARBURETOR
• IMPROVED DISTRIBUTION
• LEANER MIXTURES
• FASTER ACTING CHOKE
• EXTERNAL IDLE MIXTURE LIMITER
• SOLENOID THROTTLE STOP

CRANKCASE
VENTILATOR VALVE

HEATED
INTAKE AIR

REDUCED COMPRESSION RATIOS

INTAKE MANIFOLD

Chrysler pioneered the engine mod ification system. This is the state to which it had developed by 1970

ENGINE MODIFICATION SYSTEMS

When exhaust emission controls were in the planning stage, every car maker except Chrysler Corporation felt that the only way to control exhaust emissions was with the air pump system. Chrysler was doing a lot of experimenting and running test fleets with engine modification systems. By running leaner fuel mixtures, retarded spark, higher idle speed and different combustion chambers, Chrysler thought that they could pass the emission tests without an air pump. When the 1966 California cars rolled off the end of the assembly line, every American manufacturer had an air pump on his engine except Chrysler. Their faith in the engine modification system paid off. One year later, other manufacturers also started to use the engine modification system and the air pump was on the way out.

Air pumps were used less and less by all manufacturers as they relied on the engine modification system until 1972. In that year, to meet the stricter requirements, Chrysler finally had to put its first air pump on California engines. As the requirements get stricter, it becomes more and more difficult to make an engine pass the emissions test without the pump, but the manufacturers are doing their best to lessen emissions with engine modifications only.

Chrysler's Clean Air Package (CAP) and Clean Air System (CAS)

Chrysler's system was originally called the Clean Air Package and later the name was changed to Clean Air System. The engine modifications in the system are really very simple. A normal, uncontrolled engine pollutes very badly at idle because the mixture is so rich. The rich mixture is what causes unburned hydrocarbons and carbon monoxide to come out the tailpipe. If you run a very lean mixture the engine won't idle unless you open the throttle more to keep it running. With the throttle opened further the engine idles too fast. The easiest way to slow it down without affecting the throttle opening or the mixture is to retard the spark. This is the combination that is used at curb idle on all Chrysler engines with the Clean Air System. The idle mixture is lean, the throttle is opened further than it would be normally and the spark is retarded. The retarded spark not only helps slow the engine down, but it also increases the temperature in the exhaust manifold which oxidizes the carbon monoxide and also burns up the hydrocarbons.

To ensure that the mixture will stay lean and not be affected by somebody fiddling with the idle mixture screws, Chrysler has used several different systems to limit the amount of fuel that you can get at idle. The cleverest one of these is an idle mixture limiter built into the carburetor. No matter how far out you unscrew the idle needle, the limiter prevents rich mixtures beyond a certain point. Some carburetors have a pinned needle. If you try to unscrew the mixture needle too far, the pin breaks the needle off and then you have the problem of either drilling it out and fixing the throttle body or replacing the carburetor.

The fuel mixture is also set very lean at acceleration and cruising so that the emissions will be as low as possible. It wasn't necessary to lean it out too much in the cruising range because engines normally do not pollute very much when they are cruising at part-throttle.

The lean mixtures and other changes mentioned so far resulted in, originally, over 50% reduction in hydrocarbon and carbon monoxide emissions during idle and acceleration, but further controls were necessary to control emissions during deceleration on manual transmission cars.

When an engine without emission controls decelerates, it pollutes very badly. The rich idle mixture is sucked into the engine under high vacuum. Very little air comes into the engine because the throttle is closed. Also, the high vacuum pulls the exhaust back through the open exhaust valves and makes a very bad mixture that doesn't burn up. The wider throttle setting used with the Clean Air Package helps to lower emissions on deceleration because it lets more air into the combustion chamber. This reduces the deceleration vacuum and there is less tendency for the exhaust to be pulled back through the open exhaust valves.

Cars with manual transmissions maintain the high vacuum of deceleration much longer than a car with an automatic. The automatic gets an initial high vacuum when the driver takes his foot off the throttle, but this goes away pretty fast because of the slippage in the transmission. The automatic transmission CAP cars did not require any additional controls in most cases, but the manual transmission cars were still polluting heavily during deceleration. Chrysler engineers discovered that the best way to burn up those hydrocarbons on man-

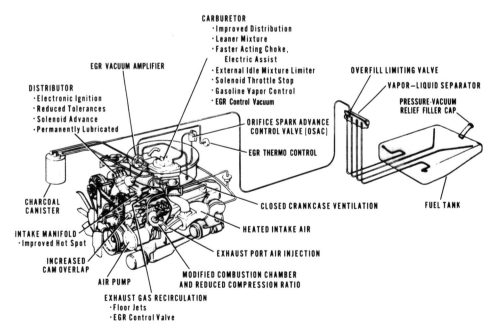

By 1973, Chrysler's original Clean Air Package had evolved to this Cleaner Air System. It's not as complicated as it looks, because you only work on one part of it at a time

ual transmission cars during deceleration was to advance the spark. Normally the distributor vacuum diaphragm goes to the neutral or no advance position when the throttle is closed. If the engineers hooked up the vacuum diaphragm to the intake manifold, then they would get spark advance during deceleration but they would also have spark advance at idle when they definitely didn't want it. The solution to this problem was the Distributor Vacuum Control Valve or "Spark Valve". The ported vacuum line from the carburetor to the distributor ran through the spark valve. Another line was connected from the valve to intake manifold vacuum. During deceleration, the high intake manifold vacuum moved the valve so that the source of vacuum for the distributor was switched from the carburetor port to the intake manifold. The spark valve would stay in this position as long as the engine was decelerating with about $21\frac{1}{2}$ in. Hg of vacuum or more. When the deceleration vacuum dropped off, the spark valve went back to its normal position and the distributor operated on ported carburetor vacuum the same as any other distributor.

Because you can't maintain deceleration vacuum on an engine that is sitting on the shop floor, the test for the spark valve is to see how many seconds it takes for it to switch back when you decelerate the engine without a load. Because the test is only for a few seconds, many people assume that the valve works that way. on the engine, only opening for a few seconds when the driver takes his foot off the throttle. Actually, the valve goes to the intake manifold position as long as the engine is decelerating with more than about $21\frac{1}{2}$ in. Hg of vacuum.

Other cars have also used the Chrysler spark valve. All of the spark valves, when used on Chrysler products or other cars, are adjustable by removing the end cap and turning the screw to increase or decrease the spring pressure. However, some car makers do not recommend adjust-

ment.

One of the internal changes in the Clean Air Package was an increased quench height in the combustion chamber. On an overhead valve engine, the quench height is the distance from the top of the piston to the underside of the cylinder head where the smooth part of the head laps over the cylinder. If this space is made very small, the mixture squirts out of the space as the piston rises. This gives a swirling to the mixture which allows a much greater compression ratio without getting any detonation or ping. Without the ping you get the power of the higher compression ratio while using lower octane gas. The quench area is bad for emissions because sometimes the fuel in that little pocket does not burn. Also, it has a tendency to collect carbon, which makes the pocket even smaller, increasing the problem. Opening up the quench area has a definite beneficial effect on emissions, but it forces the car maker to lower compression ratios.

A lot of other refinements have been made in carburetion to try to get better fuel distribution. Some of the CAP carburetors have a single idle mixture screw. This screw adjusts the mixture in both barrels of a two-barrel carburetor at the same time. Other carburetors have an adjustable idle air bleed which is adjusted at the factory.

Chrysler was one of the first to use a throttle dashpot which gave a slower closing throttle, on manual transmission cars. We normally think of the dashpot as being used to stop the engine from stalling when the driver takes his foot off the throttle on an automatic transmission car. This dashpot was not to prevent stalling, but to stop the heavy emissions caused by the rich mixture from a closed throttle during deceleration.

Another clever design Chrysler has used is their solenoid retard vacuum advance. This solenoid is part of the vacuum advance unit and it is hooked up to a ground contact on the throttle stop screw. When the engine throttle is brought back to the curb idle posi-

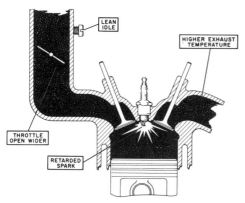

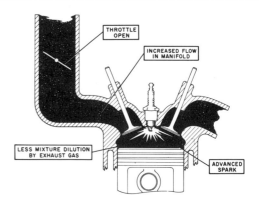

HOW CLEANER AIR SYSTEM LIMITS EMISSIONS AT IDLE

HOW CLEANER AIR SYSTEM LIMITS EMISSIONS ON DECELERATION

Spark timing and throttle opening work hand in hand. A retarded spark requires a wider throttle opening to maintain the same idle rpm. Or you could say that a wider throttle opening requires a retarded spark, for the same reason. The result is lowered emissions because of a denser mixture and a hotter exhaust

This is the result of having a spark valve that advances the spark during deceleration. When combined with the wider throttle opening that was made possible by retarding the spark at idle, there is less mixture dilution and more complete burning

tion, the stop screw touches the contact, grounds the circuit, turns on the solenoid, and the spark is retarded by a small electromagnet in the vacuum advance that pulls the breaker plate to the retard position. The advantage of the solenoid retard is that it does not operate during cold start because the throttle stop screw is held off the contact by the fast idle cam. This gives an advanced spark for better cold starting. Another advantage is that the spark retard goes away immediately as soon as the throttle is opened. This allows the engine to run more normally above idle. In 1970, Chrysler finally had to add a heated air cleaner to their system and it is now used on all of their engines.

When the Clean Air Package and Clean Air System first started, it was easy to lean over the engine compartment and point out the various components of the system. Now, so many other items have been added to the engine that it is hard to know where the Clean Air System ends and another system takes up. In 1972, Chrysler added air pumps to some California engines. They also added an NO_x spark control and exhaust gas recirculation. The old Clean Air Package is still there, but it's covered up by all the additional controls.

When repairing Chrysler Corporation cars, it is much easier if you consider each individual system on its own, without thinking of the whole thing as one big Clean Air System.

American Motors Engine Mod

After Chrysler proved that the engine modification system would work, the other manufacturers didn't waste any time in coming out with their own versions. American Motors brought out their Engine Mod system on the 232 6 cylinder engine in 1967. They used a composition cylinder head gasket which was thicker than the usual steel gasket and increased the quench height. The carburetor was set lean and the initial spark setting was retarded from 5 to 8 degrees.

In 1968, the Engine Mod system graduated to the American Motors V8s with automatic transmission. The heated air cleaner also appeared that year as the beginning of more and more units that had to be added to keep the emission levels down. In 1970, they reached the ultimate in gadgetry with a dual diaphragm vacuum advance and a deceleration valve similar to the one Chrysler was using. In 1971, transmission controlled spark took over from the dual diaphragm distributor and deceleration valve. American Motors has kept their system simple since then, sticking with transmission controlled spark on many of their cars. The reason that American Motors has not had to go to some of the complicated spark controls that other manufacturers have used, is that they were not afraid to drop an air pump on an engine when it looked as if the engine modification controls were getting out of hand. The air pump has always been the easiest solution to lowering emissions, although it is a costly one because there is so much hardware.

Late model American Motors cars seem to have just as many emission controls as most of the other cars do. Exhaust gas recirculation, transmission controlled spark, and even the air pump are still very much in the picture. Since American Motors uses the Ford carburetor they also have the Ford electric choke, which heats up the choke coil and gives a quicker choke opening. The term "Engine Mod," in reference to the American Motors engine modification system, has fallen by the wayside somewhat with the introduction of exhaust gas recirculation, transmission controlled spark, and the other systems. When working on an American Motors vehicle, it is better to consider each system individually than to try to consider the engine being controlled by one big Engine Mod system.

Ford's Improved Combustion (IMCO) System

Ford's IMCO system first appeared in the middle of 1967 on the 170 and 200 cubic inch 6 cylinder engines and on the 410 V8 in Mercury cars. None of the modifications that make up the IMCO system are visible when you lift the hood. Part of the system is the heated air cleaner, but the heated air cleaner is also used on other engines. The rest of the system consists of modifications to the carburetor, intake manifold, cylinder heads, combustion chamber, exhaust manifold, camshaft and distributor.

Modifications to the carburetor are mainly changes in the fuel flow to get leaner mixtures. Getting these leaner mixtures was not just a simple case of changing jet sizes, but involved relocating some of the components inside the carburetor such as the idle jet tube. The bowl vent designs were also changed to get better internal venting at all engine speeds.

Some Ford carburetors used an internal idle limiter needle. It was similar to the idle mixture screw, but was covered by a lead seal so that it couldn't be changed. The idle limiter needle was set to a maximum richness value so that no matter how far the idle mixture screw was unscrewed, the mixture would not richen beyond that point. Later models of Ford carburetors used idle limiter caps which are also used by most other manufacturers. Intake manifold changes were made to get better heat on the fuel passages and several of the passages were reshaped to give better distribution. Combustion chambers were changed so that they had the same volume, but less surface for unburned fuel to cling to. The exhaust manifolds were given a more free flowing design and the camshaft was changed so that both intake and exhaust valves had less open time. Both the exhaust manifold and camshaft changes reduced back flow of the exhaust into the combustion chamber at idle, thereby, giving a better mixture that would burn more completely. Initial spark timing was retarded several degrees to get more complete burning of the fuel at idle and both vacuum and centrifugal spark advance curves were tailored to reduce emissions.

As soon as it was proved that the first IMCO system would work, Ford concentrated its efforts on eliminating air pumps from as many engines as possible. To eliminate the pump, they had to go to some of the most complicated systems in the industry. Many Ford engines use a dual diaphragm distributor which retards the spark at idle. Some six-cylinder engines also use a vacuum deceleration valve so that the spark will be advanced during deceleration.

In 1970, Ford started using electronic control of vacuum spark advance. This was their Electronic Distributor Modulator System, sometimes called the "Dist-O-Vac." Vacuum advance was shut off at low speeds and allowed at high speeds. In 1972, the Electronic Distributor Modulator system was changed slightly and given the new name of Electronic Spark Control. In 1973, exhaust gas recirculation, plus a change in the Federal test procedure resulted in the elimination of the electronic spark control from passenger car engines.

Ford has had more trouble with their six-cylinder engines than their V8s. Some of the six-cylinder engines have very complicated emission control systems. Many of these systems, such as Transmission Regulated Spark, Temperature Activated Vacuum, Delay Vacuum By-pass and the Spark Delay Valve, all are used to control distributor vacuum advance. A fair question is, why didn't Ford just eliminate the vacuum advance completely? The answer is that they had to keep it for better driveability, better gas mileage, and to help prevent engine overheating.

One important part of Ford's engine modification system is the fresh air tube, which they call a "zip tube." This tube connects from the grille at the front of the car to the air cleaner and supplies the engine with cool air from outside the engine compartment. The zip tube is used to cool the mixture down and help prevent detonation. The zip tube only affects engine operation on a very hot day when the outside air temperature is around 100°F. On that kind of a day, the temperature of the air inside the engine compartment can easily go to 200°F and that is much too hot to enable the engine to breathe without detonating, under a heavy load.

Ford still uses the term IMCO to describe their engine modification system, but IMCO itself is really not a specific system on the engine. All the term means is that the engine does not have an air pump. As with the other car makers, you should consider each emission control system individually and not try to think of IMCO as being one system, covering the whole engine.

General Motors Controlled Combustion System (CCS)

Because the only visual difference in an engine with the controlled combustion system is the heated air cleaner, it looks as if that is all there is to the system.

Actually, the heated air cleaner is only one of many modifications General Motors has made to their engines when they didn't use an air pump. The original CCS system, used in 1968 and 1969, consisted of only 4 modifications. The carburetor was specially calibrated for lean mixtures and the engine idle speed was increased. The distributor was calibrated for emissions and set with initial timing that was retarded. A high temperature thermostat was used to raise the engine operating temperature and a thermostatically controlled air cleaner heated the carburetor air intake. Most of these original modifications were designed to lower hydrocarbons and carbon monoxide, but they did not do much for NO_X.

In 1970, transmission controlled spark came in on some engines. In 1973, exhaust gas recirculation (EGR) was added to many engines. General Motors still uses the term CCS as applying to all their emission controlled engines, but CCS in itself is not a specific system.

The most confusing system on General Motors cars, especially Chevrolet, is their Transmission Controlled Spark. Each year changes were made to the TCS system, in some cases making it work exactly opposite to the year before. When you work on a General Motors car with Transmission Controlled Spark, you must use only the specifications and repair procedures for that particular year. A good example of the changes in the Chevrolet Transmission Controlled Spark system are the different vacuum solenoids used. In some years the vacuum solenoid shuts off vacuum spark advance when it is energized. In other years, the solenoid is nor-

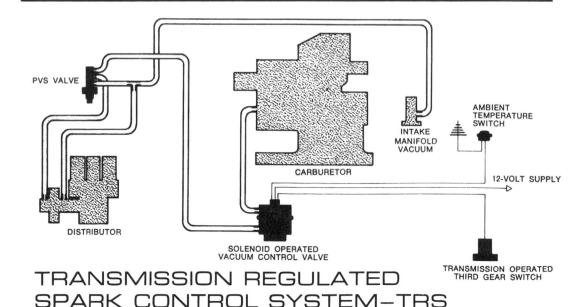

PVS VALVE

DISTRIBUTOR

INTAKE
MANIFOLD
VACUUM

CARBURETOR

AMBIENT
TEMPERATURE
SWITCH

12-VOLT SUPPLY

SOLENOID OPERATED
VACUUM CONTROL VALVE

TRANSMISSION OPERATED
THIRD GEAR SWITCH

TRANSMISSION REGULATED SPARK CONTROL SYSTEM–TRS

All transmission controlled spark or speed controlled spark systems work in the same basic way, although the hardware varies considerably. Their purpose is to cancel vacuum advance at low speeds or in lower gears

mally closed so that it shuts off the vacuum when it is not energized. You must be very careful to correctly identify the year of car you are working on, and determine that somebody has not substituted the wrong parts so that the system works backwards from the way it should.

SPARK CONTROLS

Transmission Controlled Spark (TCS)

Transmission Controlled Spark first came out in 1970 on General Motors cars, one year earlier than the law required. Transmission Controlled Spark is a system of shutting off the vacuum spark advance in the lower gears or at lower speeds. If the system is speed controlled, it is usually called speed controlled spark. However, American Motors calls their system Transmission Control Spark, although on cars equipped with automatic transmission, it is sensitive to car speed rather than gear position.

In all TCS systems, vacuum to the distributor is turned on and off by a vacuum solenoid. This solenoid receives current when the ignition switch is on and usually is fused in the car fuse block. The solenoid is

grounded at the transmission in certain gears and ungrounded in others. The difficulty in checking out the system comes from the fact that their are two kinds of solenoids, normally open to vacuum or normally closed. The normally open solenoid allows vacuum to pass through it and act on the distributor when it is *not* energized. The normally closed solenoid allows vacuum to pass through the distributor only when it *is* energized. The early systems all used normally open solenoids. When the solenoid was energized, by the transmission being in the proper gear, it closed the vacuum passage and cut off the vacuum advance. This was a fail-safe system. If the fuse blew or anything happened to break the electrical circuit, then the solenoid would de-energize and you would have vacuum advance at all times.

Some manufacturers stayed with the normally open solenoid through all the years that they used transmission controlled spark. When using a normally open solenoid without a relay, the transmission switch in the solenoid ground circuit must be normally closed. In other words, if you turn the switch on with the engine not running, the transmission switch

Spark control vacuum solenoids come in many different shapes. This is an AMC solenoid

The Pontiac vacuum solenoid used in early 1973 has a plug-in delay relay

completes the circuit and the solenoid is energized. When the transmission goes into High gear, which is usually the gear in which vacuum advance is permitted, the transmission switch is opened. This de-energizes the vacuum soleoid and allows the vacuum to pass through it and act on the distributor vacuum advance.

The CEC or Combined Emission Control solenoid used on Chevrolet and also on the General Motors 6 cylinder engine, is a normally closed solenoid used with a normally open transmission switch. This means that when you turn on the ignition switch with the engine not running, nothing happens because the transmission switch is open, breaking the ground circuit. The CEC valve just sits there and does nothing until the transmission goes into High gear. At that time, the transmission switch closes, which energizes the CEC valve and turns on the vacuum to the distributor. The CEC valve has a dual function, which is why it is called a combined emission control. The same plunger in the valve that turns on the vacuum to the distributor also extends and pushes the throttle linkage open for a wider closed-throttle opening when the transission is in High gear. When the throttle is held open, during deceleration, you don't get the high intake manifold vacuum that pulls so much fuel over from the idle circuit and makes such rich mixtures.

This helps emission control.

The stem of the CEC valve is adjustable, but it is never used to set curb idle because at idle it doesn't even touch the throttle linkage. There is an rpm setting for the CEC valve, but it is only necessary to adjust it if the carburetor has been overhauled or somebody has been tampering with the adjustment. The CEC valve can always be recognized because there are vacuum hoses connected to it. The normal anti-dieseling solenoid that you find in the same position on many carburetors does not have vacuum hoses. It has only a single wire. Curb idle is always set with the anti-dieseling solenoid adjustment. Curb idle is never set with the stem of the CEC valve.

An additional control on many transmission controlled spark systems is an ambient temperature switch or a coolant temperature switch. Temperature switches usually provide a ground at low temperatures to turn the solenoid on, and open at high temperatures so that the system is not affected.

With a normally open vacuum solenoid, the solenoid is energized at all times, except in High gear. The temperature switch provides a ground at cold temperatures, but there isn't any way that this ground can be directly hooked up to turn the vacuum solenoid off and allow vacuum advance.

The basic test of any vacuum spark control is to disconnect the hose at the distributor and connect a vacuum gauge. If the spark control requires rotating the rear wheels, you can use a long hose and position the gauge in the driver's seat while you go around the block

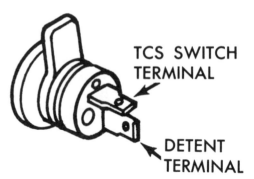

TCS SWITCH TERMINAL

DETENT TERMINAL

The big General Motors hydramatic has the transmission switch inside, with these two terminals outside

To do this, a relay is inserted into the hot wire between the ignition switch and the vacuum solenoid. The temperature control switch provides a ground for the relay at low temperatures. This energizes the relay which then breaks the circuit between the battery and the vacuum solenoid, de-energizes the solenoid, and allows vacuum advance. The relay in this type of a system would be a normally closed relay. In other words, it is closed and allows current to flow to the vacuum solenoid, except when the temperature switch provides a ground to energize the relay.

Some temperature switches also provide a hot override. When the engine temperature gets up to the danger point, the hot override operates the solenoid, sometimes through a relay, to allow vacuum advance.

Additional controls are a time relay and a delay relay. The terminology on these two relays is confusing because they have both been called delay relays. In this book, the time relay is the one that allows vacuum advance for approximately 20 seconds after the ignition key is turned

on and the engine started. The delay relay is the one that delays the application of vacuum to the distributor for 20 seconds after the transmission goes into High gear.

The time relay used on Chevrolet vehicles is always mounted on the firewall or somewhere in the engine compartment. The delay relay on Chevrolets is always mounted behind the instrument panel, so there is no chance of getting the two mixed up.

The Pontiac Transmission Controlled Spark used on 1973 models built before March 15th, is a completely different system from the Chevrolet and other General Motors systems. Some of the Pontiac spark controls also operate the exhaust gas recirculation valve. Mounted next to the Pontiac vacuum solenoid is a black delay relay, that delays the application of vacuum advance after the transmission goes into High gear. This delay relay is an innocent looking piece of equipment, but its cost is in the $20.00 range.

After March 15, 1973, Pontiac did away with the delay relay and used a start-up relay which is the same unit as the time relay used by Chevrolet.

Speed Controlled Spark (SCS)

This system uses a vacuum solenoid similar to Transmission Controlled Spark. The difference is that speed controlled spark turns the solenoid off to allow vacuum advance at about 35 miles per hour instead of when the transmission goes into a certain gear. The solenoid receives current from the ignition switch and is grounded at the speed switch in the speedometer cable. Below 35 mph, the solenoid is grounded, which energizes it, shutting off the vacuum to the distributor. The small governor inside the speed switch senses the car speed and breaks the contact with ground at approximately 35 mph. This turns the solenoid off and allows vacuum advance. American Motors uses the term "Transmission Controlled Spark" even though, on their automatic transmissions, the spark is controlled by the speed of the car rather than the gear position. Before March 15, 1973, American Motors used an ambient temperature switch at the front of the car. This switch turned off the current to the solenoid below 63°F and allowed vacuum advance at all speeds. After March 15, 1973, American Motors dropped the temperature switch and used a speed controlled spark system that works by governor pressure from the automatic transmission. Since governor pressure is directly related to speed, it amounts to the same thing as using a speed switch. Spark advance is allowed above a governor pressure corresponding to approximately 35 mph.

Temperature Spark Controls

Temperature switches are used in several of the spark control systems to cancel the system under cold conditions and allow full vacuum advance so that the engine has better driveability. Some of the temperature switches sense outside air temperature. They are located either at the front of the car at the air intake, in

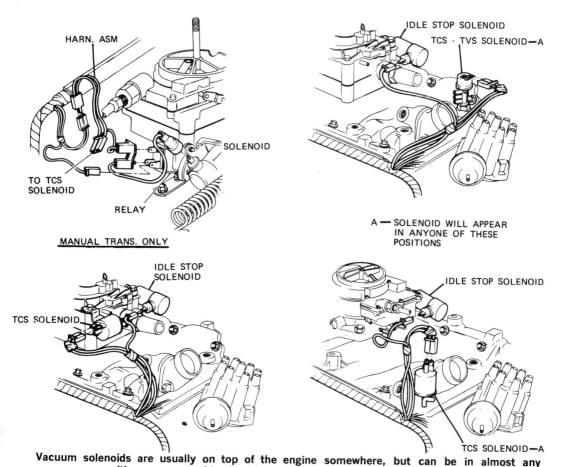

Vacuum solenoids are usually on top of the engine somewhere, but can be in almost any position, even combined with a thermo-vaccum switch, as on Buick.

All of the relays that come with the transmission controlled spark systems on General Motors cars can be tested on the bench, with hookups similar to this

the cowl, or in one of the front door posts. The switches that are sensitive to outside air temperature were eliminated after March 15, 1973 or moved to locations on the engine, so that they are sensitive to engine temperature instead of outside air temperature. To make the switches more sensitive to engine temperature, some of them were enclosed in plastic or metal housings that bolted to the engine. Others were relocated in the air cleaner. In some, the design of the switch was completely changed so that it became a coolant temperature switch that screwed into the block.

Coolant temperature switches have been used for several years to control vacuum spark advance. A coolant temperature switch is usually mechanical. A heat sensitive element inside the switch expands and changes the routing of the vacuum from one nozzle on the switch to another.

Air temperature switches can be either mechanical or electric. The electric ones usually control the flow of electricity to a vacuum solenoid or are in control of the ground circuit from the solenoid.

Another type of temperature switch, used by Pontiac, is screwed into an intake manifold passageway and senses the temperature of the air fuel mixture. The switch is the mechanical type and works much the

same way as a coolant temperature switch would. Pontiac also uses a switch that screws into the cylinder head and senses the temperature of the metal.

Another coolant temperature switch, that has 3 nozzles and has been used for many years on many different makes of cars, is called the Thermo-Vacuum Switch, Distributor Vacuum Control Switch, or the Ported Vacuum Switch. This switch is connected to intake manifold vacuum, to ported vacuum, and to the distributor vacuum advance. When used on all cars except American Motors, the switch only works when the engine is overheated. At that time, the ported vacuum from the carburetor is shut off and full manifold vacuum is sent to the distributor.

American Motors uses a similar 3-nozzle switch, but the way it works is entirely different. This switch sends intake manifold vacuum to the distributor when the engine is cold and ported vacuum when the engine is at normal operating temperatures. American Motors also has a 3-nozzle switch used with their exhaust gas recirculation system. One of the nozzles is not used and the switch is colored black for identification.

On Ford Motor Company vehicles, the 3-nozzle switch can be hooked up two different ways. The normal way

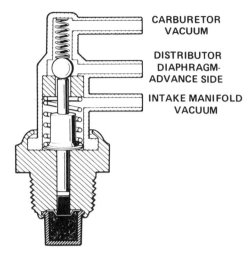

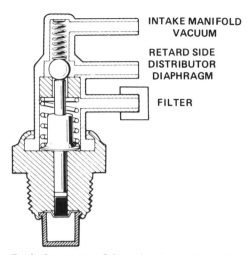

Ford calls this a ported vacuum switch. General Motors names it a Thermo Vacuum Switch. Whatever you call it, it is not an emission control, but simply a device to help cure overheating. This is the way it is hooked up on Fords when the 3-hose hookup is used

Ford also uses a 2 hose hookup with a filter, called a "Retard-Vent" system. It is used to cancel dual diaphragm distributor retard at idle

is with 3 different hoses to it, so that it sends either manifold vacuum or ported vacuum to the distributor vacuum advance. The other way is to use the switch with a dual diaphragm distributor and connect it so that manifold vacuum to the retard side of the diaphragm runs through the switch. When the engine overheats, the switch cuts off the manifold vac-

uum to the retard side and vents the diaphragm to atmosphere through the third nozzle which is covered with a little foam filter. Many of the air temperature and coolant temperature switches look the same, but inside they are entirely different and they operate differently. Always go by part number and not appearance when installing a new switch.

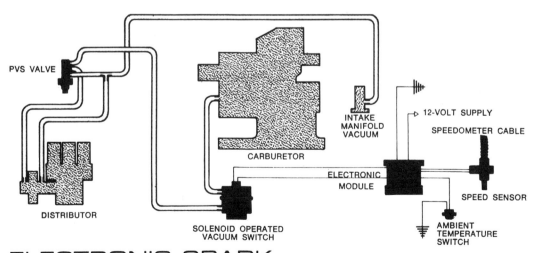

ELECTRONIC SPARK CONTROL SYSTEM—ESC

Ford's Electronic Spark Control is really nothing more than a speed controlled spark system with an electronic box to turn the vacuum solenoid on and off

Deceleration Spark Valves

The spark valve was pioneered by Chrysler Corporation, and is used mainly on its manual transmission vehicles. With a ported spark setup, the distributor vacuum advance goes to the neutral or no advance position at idle and also during deceleration. Manual transmission vehicles have a higher vacuum during deceleration because there is no slippage in the transmission. This higher vacuum draws in a lot more fuel and creates a rich mixture that has to be burned up in order to lower emissions. This mixture will burn better if the spark can be advanced during deceleration. The problem is that you do not want the spark advanced at idle. The spark valve does this job very neatly by allowing normal ported vacuum advance at all times except during deceleration. Then it switches over so that full manifold vacuum is sent to the distributor vacuum advance. The switching point for the spark valve is approximately 21 in. Hg of vacuum. As long as manifold vacuum stays above that value, the spark valve sends full manifold vacuum to the distributor which stays in t he advance position. The length of time that the valve stays in the advance position depends on the weight of the car, the incline of the road, and the speed.

Spark valves usually are tested on the shop floor by revving the engine and allowing it to decelerate with the transmission in Neutral. With a vacuum gauge at the distributor, you can see the change in vacuum and the number of seconds elapsed before the change—this is the measure of whether the valve is adjusted correctly. Some manufacturers use a different system of adjusting the valve and others do not allow any adjustment at all. They adjust the valve at the factory and if there is anything wrong with it, you are supposed to replace it.

Electronic Spark Control (ESC)

Electronic spark control is used only by Ford Motor Company. They

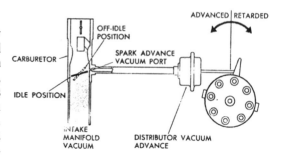

When the vacuum spark advance is "ported" it means the port is above the throttle plate so there is no advance at idle

have used two systems and both of them are similar. In 1970 and 1971, they used the Electronic Distributor Modulator or "Dist-O-Vac." In 1972, this system was improved and called the "Electronic Spark Control." Actually, the systems are nothing more than speed controlled spark with an electronic control box. The electronic control is mounted behind the instrument panel, usually near the glove compartment. In some cars, the glove compartment has an open top and you can reach in and remove the electronic control from its mounting bracket, without having to crawl underneath the instrument panel.

A speed sensor in the speedometer cable is also behind the instrument panel, only a few inches in front of the speedometer head, or in the engine compartment. This speed control is actually a little AC generator. It sends signals to the electronic control which then turns a vacuum solenoid on or off to control vacuum to the distributor.

In the 1970-71 Dist-O-Vac system, the vacuum solenoid and the electronic control were in a single large box behind the instrument panel. In 1972, the two units were split up and the solenoid was mounted on the intake manifold, but the electronic control remained in the passenger compartment, behind the instrument panel. Relocating the solenoid on the engine eliminated the long vacuum hoses that had to take vacuum all the way through the firewall to the control box.

The electronic spark control systems are nothing to worry about because you are not required to repair any of the electronics. If the other units of the system check okay, but it still won't work, then the trouble has to be in the electronic control box and all you do is replace it.

A temperature switch located in the door post was used with all the electronic systems. The 1970-71 Dist-O-Vac switch is in the ground circuit and is closed when it is cold. The 1972 electronic spark control switch is in the hot circuit and is closed when the temperature is warm. If you replace the switch, warm it up before you go to the trouble of installing it, to see if it is open or closed when it is supposed to be.

In spite of the complexity of the electronic systems, all they do is control distributor vacuum according to speed. It's an easy thing to hook up a vacuum gauge and drive a car to find out if they are working.

Delay Vacuum By-Pass (DVB)

On many engines, Ford uses a spark delay valve between the carburetor port and distributor. The valve is nothing more than a calibrated restriction that delays the application of ported vacuum to the distributor for a few seconds. When the engine is cold, this spark delay can cause problems with driveability. So, Ford set up a system with a vacuum solenoid, a check valve, and a door post temperature switch to by-pass the spark delay valve when the weather was cold. The door post temperature switch is in the hot circuit so that it turns the current to the solenoid on and off. The solenoid has one terminal receiving current from the temperature switch and the other grounded at the solenoid. The delay vacuum by-pass system was one of those defeat devices that the EPA said had to be removed. And so after March 15, 1973, all delay vacuum by-pass systems were deleted. When the engine calibration number on the sticker is followed by an "X", it means that the delay vacuum by-pass has been left off the engine.

Ported Vacuum Advance

The term "Ported Vacuum Advance" means that the distributor vacuum advance unit is connected to a small port in the carburetor throat, above the throttle valve, when the throttle is in the curb idle position. There is no vacuum above the throttle valve to act on the port, so at curb idle the distributor vacuum advance goes to the neutral or no advance position. When the throttle is opened, the throttle blade passes over the spark port exposing it to vacuum and then the distributor advances. This system has been in use for many years on most makes of cars. However, some cars use full manifold vacuum to the distributor vacuum advance. When the full vacuum system is used, the distributor is fully advanced at idle. Full vacuum advance at idle helps to control overheating.

If a car that was set up at the factory for ported vacuum advance has its hoses switched around by mistake so that it is connected to full manifold vacuum, the engine idle speed would be considerably higher. If you did not know that the switch in hoses had been made, you might try to slow the engine down to a normal idle speed by adjusting the speed screw on the carburetor. Then the engine probably would not pass an idle emission test. This mistake is not discovered very often, but it is something that you should be aware of.

MIXTURE CONTROLS

Heated Air Cleaners

All heated air cleaners use air from a stove surrounding the exhaust manifold. The hot air comes up through a tube to a flapper valve in the air cleaner snorkle. The valve is moveable so that the air to the carburetor can be regulated to give full hot air, full underhood air, or any mixture of the two. The valve is moved either by a temperature bulb that acts on it directly or by a vacuum motor that is controlled by a separate temperature sensor switch in the air cleaner. Vacuum to the vacuum motor runs through the sensor switch which is closed at cold temperatures. As the engine intake air starts to warm up the sensor, the bimetal blade moves to uncover a bleed hole which bleeds off vacuum to the vacuum motor. The spring in the motor then pushes the flapper valve to the underhood air position and the hot air is completely shut off.

Many cars now have a cold air tube or "zip tube" from the underhood opening on the air cleaner snorkle to a fresh air entry in the grille of the car. This way the air cleaner takes in cool outside air instead of hot underhood air. The reason for going after the cool outside air is that the air entering the carburetor must be warm, but not too hot. On a hot summer day when the outside air temperature is over 100°F, engine compartment temperature can easily go to 200°F. This superheated air is much too thin and it leans out the mixture to the point that detonation and engine damage can easily occur. The cold air tube keeps the incoming air down to a reasonable temperature and helps prevent engine damage.

Heated air cleaner temperature sensors can be checked with a thermometer, but usually it isn't necessary to be this particular. If the sensor is defective, it probably won't work at all

Idle Mixture Adjustments

In the days before emission controls, all you had to do with idle mixture screws was adjust the carburetor to the highest vacuum or to the best idle you could get and let it go at that. If you adjust a car that way today, you'll still get a good idle, but you won't pass an idle emission test. We have all heard the joke: "If an emission controlled car idles well it's probably illegal". Unfortunately, its no joke. Many surveys have been made and it has been found in some of them that emissions usually increase after a tune-up because the mechanic adjusts the mixture screws to give the car owner the best idle he can. This is not the way to adjust emission controlled cars. Drivers are going to have to learn, if they haven't already, that they cannot expect the glass smooth idle that they got in the old days.

The most scientific way to adjust idle is with a CO meter. If the manufacturer gives specifications for CO outputs at idle, this is the best way to make the adjustment.

LEAN DROP METHOD

An alternate method, and one that some manufacturers feel is the best, is the "Lean Drop" method. In the lean drop procedure, you set the idle

EMISSIONS INCREASE WITH IMPROPER SERVICE

Malfunction Item	Emission Increase	
	H.C. - PPM*	C.O. - %*
1 cylinder missing	358	-
Rich idle mixture (1-1/2 A.F. mixture)	-	.66
Low idle speed (100 rpm lower than specification) . .	35	-
Advance timing (6° from specification)	63	-
Plugged P.C.V. valve	43	.34
Choke set rich (2 diameters of choke rod)	40	.18
Heat riser valve stuck open	35	.18
*PPM - Parts per million % - Percentage		

Vacuum motors on heated air cleaners can be checked with a hand operated vacuum pump

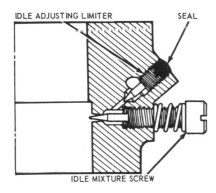

IDLE ADJUSTING LIMITER   SEAL

IDLE MIXTURE SCREW

Various means have been used to limit the amount that you can richen the idle mixture. This sealed limiter screw on a Carter YF limits the maximum amount of mixture, no matter how far you back off the mixture adjustment screw

The most common idle mixture limiter is the limiter cap. When you buy new caps from Ford, they come in this jar, submerged in fluid. Other manufacturers do not furnish the caps any more.

to the smoothest you can get and then turn the mixture screws in leaner until engine speed drops off by a certain amount. If the manufacturers specify the lean drop method, do not be afraid to use it because in many cases it is more accurate then using a CO tester that might be out of calibration. Many CO testers, even good ones, are out of calibration simply because they haven't been checked in months or even years.

Lean Best Idle

Another method of idle adjustment is called "Lean Best Idle". This is not a case of adjusting for the best idle you can get and forgetting it, it involves adjusting for the best idle, but making sure that the screws are in to the lean position as far as possible without losing any rpm. The best way to do this is to adjust for the best idle, then turn the screws in until there is a definite drop in rpm. Finally, turn the screws back out until the lost rpm is just regained. This way you have not sacrificed engine speed, but you know you are positively at the leanest setting you can get.

¼ Turn Rich

Another setting specified for idle mixture needles, usually on cars with

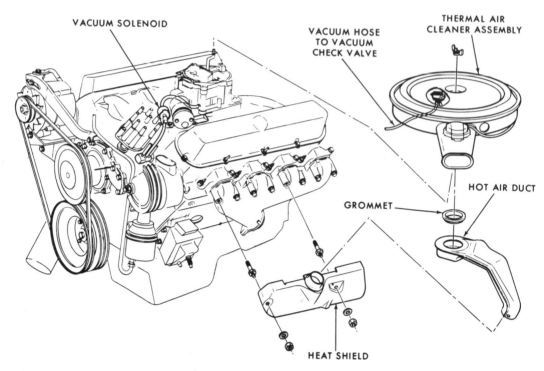

The air cleaner itself is not all there is to the heated air system. The "stove" on the exhaust mani fold is part of it

Curb idle speed is always adjusted at the anti-dieseling solenoid (lower arrow). The throttle screw (upper arrow) is only for adjusting idle with the solenoid disconnected

Sometimes, as on this Ford, the anti-dieseling solenoid has a separate adjuster, which makes it easier

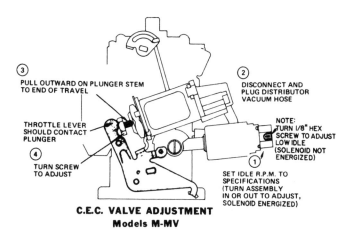

C.E.C. VALVE ADJUSTMENT
Models M-MV

③ PULL OUTWARD ON PLUNGER STEM TO END OF TRAVEL

THROTTLE LEVER SHOULD CONTACT PLUNGER

④ TURN SCREW TO ADJUST

② DISCONNECT AND PLUG DISTRIBUTOR VACUUM HOSE

NOTE: TURN 1/8" HEX SCREW TO ADJUST LOW IDLE (SOLENOID NOT ENERGIZED)

① SET IDLE R.P.M. TO SPECIFICATIONS (TURN ASSEMBLY IN OR OUT TO ADJUST, SOLENOID ENERGIZED)

The General Motors CEC valve is not an anti-dieseling solenoid, and must never be used to set curb idle. There is a special rpm specification for adjustment of the CEC valve

an air pump, is one quarter turn rich from lean roll. Lean roll is not a very precise point so that it can be hard to find on some engines. But what it means is that you turn the mixture screws in until you get a definite fall off and roll in engine rpm. Then you back out one quarter turn from that point. This gives you a richer mixture then you would normally have. But a rich mixture is what cars with an air pump need because they have to have enough fuel passing through to the engine to keep the fire going in the exhaust manifold and burn up the hydrocarbons and CO. On all cars you must follow the car manufacturers specifications for setting idle and you must not take it for granted that because it was set one way last year, it is set the same way this year.

Throttle Positioning Devices

Chevrolet's CEC valve, besides being a vacuum solenoid that controls spark advance, also opens the throttle during deceleration in High gear. This helps prevent the rich mixtures that would be pulled through the idle circuit if the throttle were allowed to close to the normal curb idle position. Many imported cars also use throttle positioning devices that are usually vacuum controlled, but may have an electrical back-up. Toyota uses a throttle positioner that operates

Deceleration Enriching Valves

Some 4 cylinder engines go very lean during deceleration. This causes incomplete combustion and emission levels become very high. If a little extra fuel and air is fed into the intake manifold during deceleration, better combustion results and emission levels are lowered. Actually, the term "Enriching Valve" is a misnomer. It really doesn't make the mixture any richer, it just puts more mixture into the engine. Some of the valves on imported cars are built into

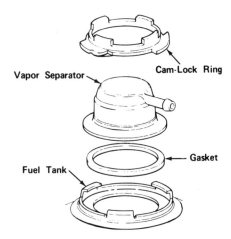

Vapor Separator

Cam-Lock Ring

Gasket

Fuel Tank

This Ford dome-type separator is held in place by a locking ring, similar to a tank gauge sending unit.

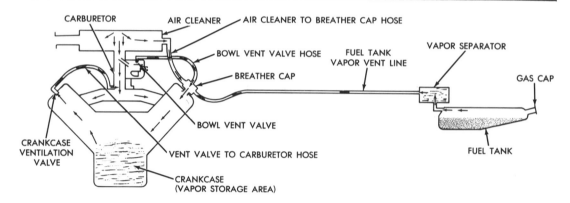

Chrysler's original vapor control system used crankcase storage. The vapor line tied in with the PCV system

the carburetor so that they feed more fuel, but not any air. However, they are used in conjunction with a throttle positioner which gives the engine extra air to mix with the extra fuel.

The separate deceleration enriching valves such as the type used on the Pinto, can sometimes develop a vacuum leak in the diaphragm. When this happens, air enters the engine at all times causing rough idle and a lean condition. If the type of valve that is built into the carburetor were to be stuck open, then, of course, there would be of an extremely rich

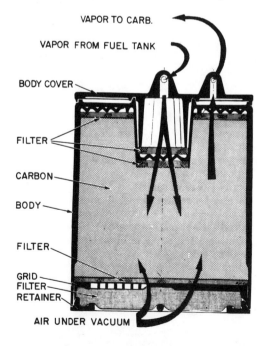

The 2 hose canister is the simplest in operation. Canisters also come in three and four hose models

mixture at all times. This is something to be on the lookout for if you have a small imported car with a high CO level.

EVAPORATIVE CONTROLS

Vapor controls are the most simple and trouble-free emission control on the whole car. They are known by various names such as American Motors Evaporative Emission Control, (EEC), General Motors Evaporative Emission Control, (EEC), or Evaporative Loss Control, (ELC), or Evaporation Control Systems, (ECS), and Evaporative Emission System, (EES). Chrysler calls theirs the Evaporation Control System, (ECS), and Vapor Saver; while Ford sticks by the term Evaporative Emission Control System, (EECS). We call all of the systems vapor control for simplification.

When vapor controls first came out in 1970, there were two ways of storing the vapors from the tank, in the engine crankcase or in a canister full of carbon (charcoal) particles. In later years all of the manufacturers went to the canister system.

Most of the vapor control system is in the tank and in the means of separating the vapors from liquid fuel. All of the tanks have some kind of fill control that keeps an air space at the top of the tank so that liquid fuel will not travel along the vent line to the vapor storage. The early systems had extremely complicated vapor liquid separators that were mounted near the tank. Several vent lines, usually

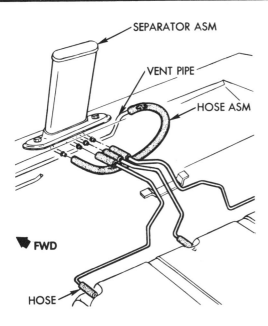

Most of the improvements in vapor control design have been in the vapor separators. The early designs used the stand pipe style separator, which is usually mounted behind the rear seat

one from each corner of the tank, led to the separator and then a single line went up to the front of the car to either the crankcase or the carbon canister. In later years, the manufacturers discovered that all that complicated plumbing was completely unnecessary. The vapor separator now

is simply a small dome on top of the tank usually filled with foam or other substance. The vapors will go through the foam, but any liquid fuel drains back into the tank. The tanks are all completely closed and use a filler cap that will open automatically if either pressure or vacuum builds up in the tank.

Vapor movement through the vapor line to the storage in the engine compartment depends upon many things. If the engine is running, fuel is being pulled out of the tank by the fuel pump and air has to enter to keep from collapsing it. This air enters through the vapor line. In effect then, the carbon canister is not only for vapor storage, but it is the point where air enters the tank anytime gasoline is being withdrawn. Air also enters through the carbon canister if the tank cools off in the evening from the heat of the day. If the carbon canister or vent line should become plugged, the safety feature is the fuel tank cap which will open either under pressure or vacuum to keep from collapsing the tank. Many of the carbon canisters have a filter at the bottom that can be slipped out and replaced on a mileage or time basis. Ford uses a sealed canister. The small-size canisters that were used on earlier models are replaced

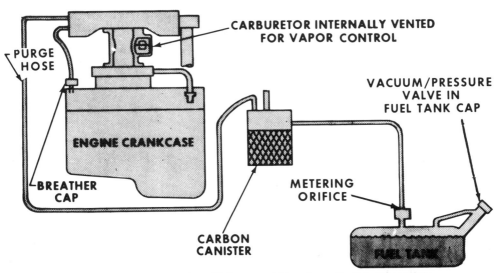

On General Motors, American Motors, and Chrysler cars, the only maintenance of the vapor control system is replacement of the canister filter, as shown here

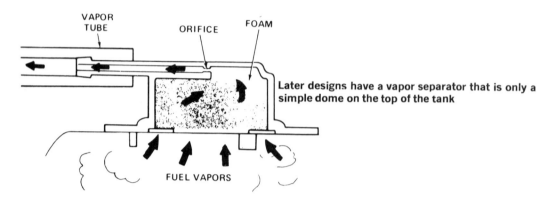

Later designs have a vapor separator that is only a simple dome on the top of the tank

completely on a mileage or time basis. The large-size Ford canisters that are now in use are good for the life of the car and there is no maintenance necessary. Undoubtedly the biggest problem you will ever have with any vapor control system is leakage from deteriorating hoses.

NOx CONTROLS

Spark Retard

The transmission controlled spark and speed controlled spark systems that first appeared in 1971 were put on the cars to control NO_x emissions. NO_x forms in the combustion chamber in excessive amounts when the peak combustion temperature gets over 2,500°F. Peak combustion temperature is directly related to spark timing. If the spark occurs at exactly the right instant, you get a maximum amount of pressure and heat in the combustion chamber and the car puts out maximum power. If the spark is retarded slightly, the power falls off because not as much heat is being generated in the combustion chamber. Keeping the heat down by retarding the spark lowers the formation of NO_x to the levels that were required in 1970-1972.

The transmission controlled spark and speed controlled spark systems are covered elsewhere under those headings. Chrysler Corporation did not refer to their system in those terms, but simply called it the NO_x control system. The system was only used two years, 1971 and 1972. Manual transmissions were set up fairly simply, allowing vacuum spark ad-

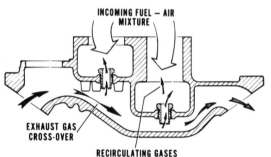

The simplest form of exhaust gas recirculation you can have is this floor jet system, used on some Chrysler products. It recirculates at all times during engine operation

vance in High gear only. A solenoid was energized in the lower gears to shut off the vacuum to the distributor. In High gear, a transmission switch opened the ground circuit from the solenoid, de-energizing it and allowing vacuum to advance the spark. The solenoid was mounted at the back of the engine, near the distributor. A temperature switch, mounted on the firewall, sensed the temperature in the plenum chamber at the cowl air intake. The temperature switch was also in the ground circuit from the solenoid. When air temperatures were cold, the temperature switch broke the circuit to ground so that the solenoid was de-energized, allowing vacuum advance. The temperature switch was left off the 1972 manual transmission cars, so that the system consisted simply of a vacuum solenoid and a transmission switch.

The system on the automatic transmission was much more complicated. In 1971, there was a vacuum control, an electronic control box, a thermo switch, a speed switch and a vacuum solenoid. The vacuum switch, the speed switch, and the temperature switch all provided separate grounds for the electronic control. These grounds were reversed in their effects on the vacuum solenoid by the electronic control box. When any one of the 3 units were grounded, the electronic control turned the solenoid *off*, allowing vacuum advance. To cut off vacuum advance, all 3 of the switches had to be opened. The temperature switch was opened above 70°F. The vacuum switch was opened below 15 in. Hg of intake manifold vacuum, and the speed switch was opened below 30 mph. The only time all 3 switches were opened was in warm weather, below 30 mph, when the engine was accelerated fast enough to drop the intake manifold vacuum below 15 in. Hg. This was quite an ingenious method of controlling vacuum spark advance because this gave advance at all times except when a cutoff of the advance was needed to control emissions.

The temperature switch, vacuum switch, and electronic control were mounted on the firewall and called the control unit. The temperature part of the control unit stuck through the firewall into the front of the cowl chamber to allow ambient temperatures to act on it.

In 1972, the vacuum part of the control unit was discontinued. This simplified the system greatly and meant that there was vacuum advance above 30 mph in warm weather, but never below 30 mph. One thing you must remember about the electric circuitry in the NO_x system is that grounding the electronic control ground circuit turns the solenoid off.

Fewer cars used the transmission control spark system after 1972 because the Federal test procedures were changed and exhaust gas recirculation came in which did a much better job of controlling emissions of NO_x.

Exhaust Gas Recirculation (EGR)

Exhaust gas recirculation is used primarily to lower peak combustion temperatures and control the formation of NO_x. NO_x emission at low combustion temperatures are not bad, but when the temperature goes over 2,500°F, the production of NO_x in the combustion chamber shoots way up. You can cut down on the peak combustion temperatures by retarding the spark or by introducing an inert gas to dilute the fuel air mixture. Introducing exhaust gases into the combustion chamber is a little like throw-

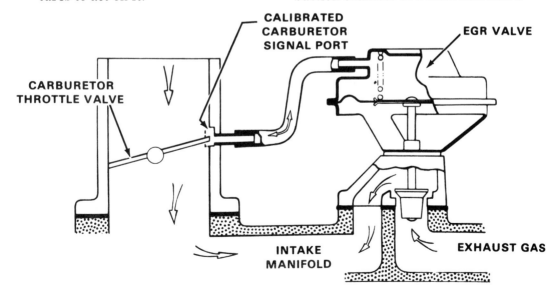

Opening the throttle will make the **EGR** valve move up and down while the engine is **idling**, and you can feel the movement of the diaphragm by putting your fingers under the valve

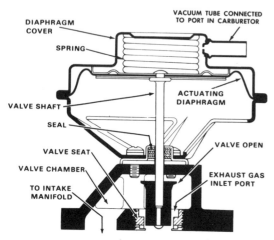

DIAPHRAGM COVER

VACUUM TUBE CONNECTED TO PORT IN CARBURETOR

SPRING

ACTUATING DIAPHRAGM

VALVE SHAFT

SEAL

VALVE SEAT

VALVE OPEN

VALVE CHAMBER

TO INTAKE MANIFOLD

EXHAUST GAS INLET PORT

Cutaway of a typical General Motors EGR valve. The Chrysler and Ford valves are similar

Testing for vacuum at the EGR valve is easily done with a gauge, as on this Ford

ing water-soaked wood on a roaring fire. The water-soaked wood won't burn, so that the fire cools down and doesn't roar nearly as much as it used to. Put a little exhaust gas in the combustion chamber and it takes the place of a certain amount of air fuel mixture. When the spark ignites the mixture, there isn't as much to burn so that the fire is not as hot. Also, the engine doesn't put out as much power.

Exhaust gas recirculation is kept to very low limits. The hole in the EGR valve is very small even when it is wide-open. It's surprising how little exhaust gas it takes to cool down the peak combustion temperatures.

Chrysler had one of the simplest exhaust recirculation systems with their floor jet under the carburetor. Holes were drilled in the bottom of the intake manifold and calibrated jets were screwed into the holes. The holes penetrated into the exhaust crossover passage and allowed the exhaust to come into the intake manifold at all times. The trouble with that system is that it allowed exhaust gas recirculation at idle, which wasn't necessary and didn't make for the smoothest idling engines. Floor jet EGR is still used on some Chrysler engines, but most of them now use a separate EGR valve the same as all the other manufacturers.

The EGR valve is mounted on the intake manifold so that when it opens, exhaust gases are allowed to go from the crossover passage into the throat under the carburetor. The EGR valve is vacuum-operated, sometimes by intake manifold vacuum and on some engines by ported vacuum. The ported vacuum systems are the simplest. At idle, the port is above the throttle blade, so that the EGR valve stays closed. When the throttle is opened, vacuum acts on the port and the EGR valve opens. At wide-open throttle, there is no intake manifold vacuum so that the EGR valve closes to give the engine maximum power.

Some cars operate their EGR valve from intake manifold vacuum. They use an amplifier in the circuit to turn on the vacuum to the valve. The amplifier is controlled by venturi vacuum. A small hole in the carburetor venturi picks up vacuum when the airflow through the carburetor is high enough and sends the vacuum signal to the amplifier. The amplifier then opens to allow manifold vacuum to act on the EGR valve. The amplifier system is used to obtain precise control over when the EGR valve operates. Also, it means that exhaust recirculation does not start until the engine is considerably above idle.

Most of the EGR systems use a

Checking the EGR valve for a leaking diaphragm is best done with a hand vacuum pump

temperature control of some kind. This can be electric or strictly mechanical. American Motors used a unique system of two air bleeds called modulators. These were mounted behind the grille and on the firewall. At low temperatures, one of the modulators would open and at high temperatures the other opened, allowing air to bleed into the vacuum system to the EGR valve so that the vacuum was weakened and the valve did not open as much. Chrysler used a similar system with an air bleed that sensed temperature inside the cowl plenum chamber. These systems that were sensitive to outside air temperatures were discontinued after March 15, 1973 as a result of the EPA order.

Ford uses a temperature control that looks like a PVS valve, but only has two nozzles. It shuts off the vacuum to the EGR valve at low temperatures. When Chrysler dropped their air temperature sensor in the plenum chamber, they went to a valve similar to Ford's, but mounted in the radiator. The valve has two nozzles with a hose connected to one and a foam filter on the other. At low temperatures, the valve opens which allows air to enter and weakens the vacuum so that the EGR valve stays closed.

Buick has changed their EGR temperature regulation considerably. In 1972, they didn't use any temperature control at all. In 1973, they had a temperature switch in the hose that shut off the vacuum to the EGR valve at low temperatures. This switch was sensitive to engine compartment temperature and was judged a defeat device by EPA, so that on March 15, 1973, Buick changed the switch to a coolant temperature switch working with a vacuum solenoid. At low temperatures, the coolant switch operated the solenoid to shut off the vacuum to the EGR valve. In 1974, Buick got rid of the electrics in the system and went to a straight coolant-vacuum switch that closed off the vacuum to the EGR valve at low temperatures.

Oldsmobile and Cadillac used a switch in the hose similar to Buick's first switch. After March 15, 1973, they enclosed the switch in a housing so that it was more sensitive to engine temperature rather than underhood temperature.

Pontiac probably has the most complicated system of all. Before March 15, 1973, the EGR system was tied in with the transmission control spark system. The two systems were hooked

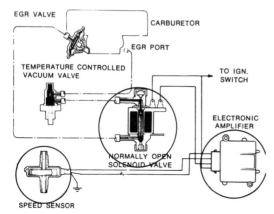

Ford had the ultimate in EGR valve control, an electronic "black box" that turned the exhaust gas recirculation off above 64 mph

together so that when vacuum spark advance was allowed, there was no EGR. When EGR was allowed, there was no vacuum spark advance. This complicated system was eliminated on March 15, 1973 and from then on the EGR and the transmission control spark systems were separate.

IMPORTED CAR EMISSION CONTROL SYSTEMS

Imported car emission control systems accomplish the same ends as the domestic systems, but usually they do it in a different way. In their crankcase ventilation systems, they sometimes use a valve the same as the domestic systems, but in other instances there is nothing but a hose connecting the crankcase to the air cleaner. This accomplishes the same thing as the valve type systems because the slight suction in the air cleaner draws the fumes out of the crankcase to keep them from going into the atmosphere.

Most imported cars use a heated air cleaner, sometimes with vacuum control to the air valve and sometimes with a simple thermostatic bulb to open and close the valve. Some imported cars use a simple door on their air cleaners which is operated by a lever on the air cleaner snorkle. The lever has two positions "summer" and "winter," which must be changed manually, depending upon the season.

The imported car air pump systems pump air into the exhaust manifold near the exhaust valve, the same as on domestic systems. There are still quite a few imported cars that use an antibackfire valve which allows pumped air to go into the intake manifold during deceleration. Other imported cars use the more modern diverter valve which vents all of the air from the pump into the atmosphere through a silencer.

Imported cars have made many modifications to their carburetors. They may use fuel cutoff valves that cut out the idle circuit during deceleration to eliminate the rich mixtures. There are also fuel cutoff valves that shut off the idle circuit when the ignition switch is turned off to prevent dieseling.

Throttle positioners and deceleration enrichers are common on imported engines. The throttle positioners open the throttle during deceleration, but sometimes have additional controls that put the positioner into operation in certain gears or above certain speeds.

Deceleration enrichers are built into many Japanese carburetors. During the high vacuum of deceleration, an extra passage opens and allows more fuel to enter underneath the throttle blade. These systems are called the "coasting richer" by many Japanese makers, probably because that was the closest translation they could get to deceleration enricher.

Spark control systems include both transmission controlled spark and a dual point retard system used by Datsun. The dual point system is a clever way of getting a few degrees of retard whenever it is needed to control emissions. The retard breaker points are placed 7 crankshaft degrees away from the normal breaker points in the distributor. A relay is used that can switch the coil primary from the normal to the retarded set of points. Anytime the retard is needed, all that is necessary is to close the switch to activate the relay and the engine is automatically 7 crankshaft degrees retarded. Depending on the year of the car and whether it has an automatic or manual transmission, several different kinds of switches have been used to control the relay. You may find a temperature switch, a throttle position switch, a transmission gear position switch and a second throttle position switch. One of the throttle position switches works when the throttle is closed and the other when the throttle goes to the wide-open position. When used with the transmission gear switch, you might call the system transmission controlled spark, but the difference is that it is not controlling distributor vacuum advance, but a simple mechanical retard of 7 degrees.

Exhaust gas recirculation is also used on many imported vehicles. In some cases they use a valve similar to the domestic system and in others they have a simple tube coming from the exhaust manifold to the intake manifold with a restrictor to limit the amount of exhaust gas that is recirculated.

Federal or State Approved Accessory Systems

This section describes the exhaust emission control devices that are required on certain years and models of used cars in California. Up to the end of 1973, California was the only state requiring installation of any aftermarket or used car emission control devices. If you live in California, the need for this section is clear. If you are in another state of the Union, there are two ways in which you may

All domestic cars now use the canister for storing vapors

come in contact with one of these devices. It could be installed on a car that was formerly used in California or it is possible that your state will eventually pass a law requiring these same devices on their used cars. Either way, the following information will prepare you for working on the devices if you come in contact with them.

DEVICES FOR 1955-65 CARS

Control devices for exhaust emissions were not installed on new cars until 1966. Because California's smog problem is acute, especially in the Riverside area, the legislature passed a law requiring that 1955-65 cars in critical smog areas must have an exhaust control device installed on transfer of ownership. There is a lot more to the law than just that simple statement. Some cars are exempt because of engine design, but we won't go into that here.

Two devices have been accredited for installation, the GM device and the Pure Power device. Because they are only accredited for use on 1955-65 cars, they are known as the 1955-65 devices. Any manufacturer may submit a device for accreditation. If the device passes, it may be sold and installed to satisfy the law. The main limitation, aside from the requirement that the device must lower emissions to a specified level, is that the installed price of the device must not exceed $85.00.

Both of the 1955-65 devices lower emissions by controlling vacuum spark advance. Other manufacturers are submitting devices, and it's possible that there may be some devices accredited later on that use other means of lowering emissions.

General Motors (GM) Device

The GM kit is installed on 1955-1965 cars. When installing the kit, the vacuum spark advance is disconnected from the carburetor, and connected to a valve mounted in the upper radiator hose. The other side of the valve is connected to intake manifold vacuum. With this system, the vacuum supply to the distributor is shut off at all times except when the engine overheats. If overheating occurs, the valve in the upper radiator hose senses the increased temperature and sends full manifold vacuum to the vacuum advance unit in the distributor. This increases idle speed and helps cool the engine. On air-cooled engines, such as the Corvair, a special kit is used, which does not have the coolant valve. Because the effect of the GM kit on engine operation is simply a disconnection of the vacuum advance, many people wonder why California didn't just require that all vacuum advances be disconnected and let it go at that. The problem was that they had to make sure that disconnecting the vacuum advance would not damage an engine.

The hardware in the GM kit is there just for that purpose, as a safety measure in case the engine overheats. The GM kit has the advantage of low cost (about $15.00 installed) and being such a simple design that there is hardly anything that can go wrong with it.

Pure Power Device

Pure Power is also used on 1955-1965 cars. This device is considerably more complicated than the GM device. The heart of the Pure Power system is an electronic control box which mounts on the fender panel inside the engine compartment. This control box controls both vacuum and initial advance. It also contains a capacitor discharge ignition system. The CD ignition system was added because the makers felt it would fire fouled spark plugs better and thereby reduce emission levels. The vacuum advance control shuts off vacuum to the distributor between 900 and 1,700 rpm. Above 1,700 rpm, the engine gets all the vacuum advance that was originally built into it. Initial advance is controlled by a small timing module that is selected according to the amount of advance on the engine and then plugged into the back of the Pure Power control unit. At idle, the spark is electronically retarded so that it occurs at zero degrees advance or Top Dead Center. As the engine speeds up, the initial advance is allowed to come back in until at approximately 1,700 rpm, the advance is completely back in.

GM EMISSION CONTROL SYSTEM FOR USED CARS

IGNITION TIMING SET

THERMO VACUUM SWITCH FOR OVER TEMPERATURE CONTROL

IDLE SPEED INCREASED AND SEALED

VACUUM ADVANCE INOPERATIVE DURING NORMAL OPERATION

IDLE MIXTURE ADJUSTED LEANER AND SEALED

The GM kit is simply a disconnection of the vacuum, advance, and installation of a valve that sends full manifold vacuum to the distributor if the coolant temperature at the upper hose goes above 205°F.

DEVICES FOR 1966-70 CARS

Engine modifications that lowered HC and CO emissions on 1966-70 cars made NO_x emissions worse. The car makers relied heavily on lean mixtures, which raised combustion temperatures above 2,500°F and increased the formation of NO_x. California realized that it was being smothered by the increased NO_x from the 1966-70 cars, so a law that NO_x devices had to be installed on those cars without NO_x control was passed. This includes all cars 1966-70, except some of the 1970 General Motors cars, which had transmission controlled spark as an NO_x control. In 1971 and thereafter, all cars had some kind of NO_x control.

At the end of 1973, six devices met the requirements for the 1966-70 law. Some cars will accept any one of the six devices. Other cars are limited in choice due to their engine size or other design features allows them to use only one of the devices. California law states that the devices must be installed for not more than $35.00. Following are descriptions of the six devices.

STP Pollution Control Device

STP's device for 1966-70 cars relies mainly on exhaust gas recirculation to lower NO_x emission levels. The device is installed by drilling a hole through the heat riser part of the intake manifold to pick up the exhaust gases. Exhaust gases flow to an EGR valve, where they are mixed with fresh air from a hose connected to the clean side of the air cleaner. The mixture of exhaust gas and clean air is fed through a hose to the intake manifold with the connection usually being made at the PCV vacuum fitting. The EGR valve mixes both the exhaust gases and the fresh air. Exhaust gas is recirculated during idle, acceleration and cruising, but not during deceleration. Fresh air is admitted during idle, deceleration and low speed steady cruising. During acceleration and high speed steady cruising, frseh air is shut off.

An important part of the EGR system is a distributor vacuum bleed or spark delay valve. The earlier kits used the vacuum bleed which allowed a certain amount of air into the vacuum advance hose and reduced the amount of vacuum advance slightly. The later kits use a spark delay valve which gives the same amount of vacuum advance, but lets it come in slower.

Echlin Device

The Echlin device consists primarily of disconnecting the vacuum spark advance and installing a plate under the carburetor that admits air through what Echlin calls a "sonic energizer". The energizer is attached to the side of the air cleaner, because that is a convenient mounting place, and connected to the plate under the carburetor by a hose. In operation, the energizer sounds just like an air bleed, but Echlin claims that the shape of the energizer and the special hose which they use creates sonic waves underneath the carburetor, which creates a microscopically fine, uniformly dispersed mixture for better combustion. In order to eliminate the temptation that somebody might reconnect the vacuum advance, the Echlin kit contains a breaker plate locater bracket that is installed in the distributor after the vacuum advance unit is completely removed. If the vacuum advance were removed and the bracket not installed, the breaker plate would be free to rotate back and forth with nothing holding it in position.

Perfect Circle (Retronox) Device

Retronox is manufactured by the Perfect Circle Division of Dana Corporation. Retronox relies mainly on exhaust gas recirculation to lower emission level, but it also controls vacuum spark advance at certain speeds. There are many different Retronox kits due to the different shape of mounting brackets for the EGR valve and sizes of hoses. In spite of the variety of kits, there are really only two basic designs. One with an engine speed switch and the other with a vacuum delay valve. The vacuum delay valve is used on engines which operate their distributors by ported vacuum. The system is such that both the EGR valve and the distributor vacuum advance are delayed approximately 10 to 15 seconds after vacuum first acts on the carburetor port. At curb idle or wide-open throttle there is neither vacuum advance nor exhaust gas recirculation. During all steady speed cruising there is the same amount of vacuum advance that would be present without the device and there is also exhaust gas recirculation.

The kits with the speed switch are used on those engines which operate their distributors by full intake manifold vacuum. The speed switch has a single wire that hooks to the battery side of the ignition coil. This is important; if it should ever be hooked to the distributor side of the ignition coil, the speed switch will be destroyed. Intake manifold vacuum is run to the speed switch and then both the distributor vacuum advance and the EGR valve are connected to the switch. There is no vacuum advance and no exhaust gas recirculation until the engine speed exceeds 1,300 rpm. This is approximately 26 mph on most cars. Because the EGR valve depends on intake manifold vacuum for its operation, it does not recirculate exhaust gas at wide-open throttle because intake manifold vacuum is almost nonexistent at that time. Exhaust gas does continue to recirculate during deceleration as long as engine speed does not drop below approximately 1,100 rpm. The speed switch is usually mounted on the firewall in the engine compartment and the EGR valve, which Perfect Circle calls their "Hyconox" valve, is mounted on a bracket on the intake manifold or rocker cover. The connection for exhaust gas is taken by drilling a hole in the exhaust pipe between the engine flange and the muffler. Stainless steel tubing is supplied to conduct the exhaust gases to the EGR valve.

Carter Device

The Carter device is simply an elimination of vacuum spark advance below 60 mph with a thermo protection switch which will restore vacuum advance at all speeds, if the engine overheats. Vacuum to the distributor is turned on and off by a vacuum solenoid which is controlled by an electronic speed sensor. The speed sensor connects to t he negative or distributor side of the ignition coil and senses engine rpm. The speed sensor is adjustable and Carter supplies a booklet with every kit that gives the engine rpm at which the speed sensor should be set for each car. In this way the speed sensor can be set on the shop floor without having to attach a tachometer and drive the car 60 mph.

Part of the Carter kit is adjustment of the idle mixture to lean best idle and a resetting of the initial timing advance to 4° retarded from the original specifications. For example, if the original specifications were 6° BTDC when the Carter kit is installed, the initial timing should be reset to 2° BTDC. In the Carter kits that came out before August 1973, the initial timing was Top Dead Center (TDC) for 225 cu in. and larger engines, 2° BTDC for 224 cu in. and smaller engines and the timing was to be left at the manufacturer's suggested initial setting if the 4° retarded setting occurs after top dead center (ATDC).

Kar-Kit Device

The Kar-Kit device is the simplest of all. It consists of two rubber caps, a set of instructions and 3 self-sticking labels. The large decal goes in the engine compartment and gives the new timing specifications that must be used. If the original timing specifications on the engine was 1° before top dead center (BTDC) or more retarded than that, such as top dead center (TDC) or after top dead center (ATDC), then the timing is not changed. If the original timing was more than 7½° BTDC, it is reset to half of whatever the manufacturer's original specification was. If the original specification falls between 7½° and 1° BTDC, it is reset to 1° BTDC.

There are 3 cautions that go along

with the Kar-Kit that must be followed. Number 1—Do not operate vehicles equipped with the Kar-Kit at sustained speeds above 60 mph (short periods for emergency and passing are okay). Installation of the Kar-Kit on vehicles which operate a major part of the time at either sustained high speed or heavy load conditions (such as towing a trailer) is not recommended.

Number 2—Do not install the Kar-Kit on engines of less than 140 cu in. or on engines which have distributors without centrifugal or vacuum advance. Number 3—Do not install the Kar-Kit on engines with impaired or defective cooling systems.

The two rubber caps in the Kar-Kit are used to block off the normal source of vacuum to the distributor vacuum advance. If the engine has a dual diaphragm distributor, the normal vacuum supply to the retard diaphragm is also blocked off. However, any temperature control vacuum supplied by thermal vacuum valves or PVS valves must be left connected so that the engine will get full manifold vacuum advance if it overheats. The two stickers in the kit are a notice to new owners warning them that the Kar-Kit has been installed and a notice that goes on the speedometer and warns the driver not to maintain any speed above 60 mph. The final requirement of the Kar-Kit is that the installation manual must be put in the glove compartment and remain with the vehicle at all times.

WHICH DEVICE IS BEST?

The normal question that everybody asks is which of the six 1966-70 kits is the one for me. There has been much controversy in California over these kits, with some people saying that disconnecting or modifying the vacuum spark advance will cause engine overheating and destruction of exhaust valves. Others have stated that they didn't see why they had to pay for a kit when all that was necessary was to require that every car owner disconnect his own vacuum advance and leave it that way.

The cost of each kit including installation does not exceed $35.00.

Therefore, they are all very close in price with some manufacturers offering discounts for quantity purchases. The Kar-Kit of course, is much cheaper than any other, selling in the neighborhood of $3.00. Whatever any one of the kits sell for, the shop owner is allowed to make the price come out to $35.00 if he wants to. Actually, there are 7 kits that qualify for the 1966-70 program. The Air Quality Product Pure Power device, listed above under "1955-65 Devices", has been given an exemption. 1966-70 vehicles equipped with the Pure Power device are acceptable as is and do not have to have a 1966-70 NO_x device installed. The law had to be written this way because the Pure Power device exceeds the legal maximum purchase/installation charge of $35.00.

There are many things to consider in selecting one of these devices for your cars. All of the devices require different installation methods. The ease of installation should be a big factor in whether you want to use the particular device or not. Also, the type of driving that you do should be considered. If there is a local representative of the company that makes the device in your area, he could be a great help if you ever run into any trouble. All of the kits will change gas mileage.

The effect of the individual devices on driveability or gas mileage is extremely difficult to predict. The state of California has tested the devices and published the results, but in some cases the device manufacturers do not agree with those results. About the only thing you can say for certain is that sometimes a device will not make any difference in the operation of the car, and sometimes it will. If you are looking for something to go wrong after the device is installed, then you will probably have some kind of complaint. If you are not sensitive to what is happening under the hood you will probably never notice anything different. We suggest you study the installation instructions, and then make your own decision as to which device is the best for you.

GM Device

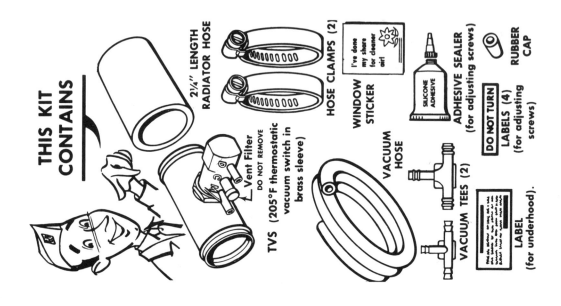

THIS KIT CONTAINS

2¼" LENGTH RADIATOR HOSE

HOSE CLAMPS (2)

Vent Filter
DO NOT REMOVE

TVS (205°F thermostatic vacuum switch in brass sleeve)

WINDOW STICKER

I've done my share for cleaner air!

SILICONE ADHESIVE

ADHESIVE SEALER
(for adjusting screws)

RUBBER CAP

DO NOT TURN LABELS (4)
(for adjusting screws)

VACUUM HOSE

VACUUM TEES (2)

LABEL
(for underhood).

GM

Delco Parts

Exhaust Emission Control

for Used Motor Vehicles under 6001 lbs. G.V.W. and Engines over 140 cu. in. displacement

Exhaust Emission Control installations are designed to reduce exhaust emissions on a motor vehicle under normal operating conditions and do not improve engine performance or economy. This kit should not be installed on a car with major engine malfunctions, such as inoperative cylinders or misfiring spark plugs, unless the owner is willing to have these malfunctions corrected. Since continued public support is an essential part of an air improvement program, the cost to the owner for installation of this kit should be kept as low as possible. The need for any work beyond the normal installation procedure should be carefully explained and only those repairs necessary to achieve normal engine operation should be recommended. Engine timing and carburetor adjustments are an important part of the installation procedure. Continued exhaust emission control is dependent on maintaining the engine in good general operating condition.

FITS MOST MAKES OF CARS AND LIGHT TRUCKS

Kit	Part No.	Fits Radiator Inlet Hose
A	3023802	1¼"
B	3023803	1½" & 1⁹⁄₁₆"
C	3023804	1⅝" & 1¾"
D	3023805	2"
E	3023806	Use for INITIAL INSTALLATION on Air Cooled Engines and for ANNUAL EMISSION CONTROL ADJUSTMENTS of all cars.

GM Device—Cont'd

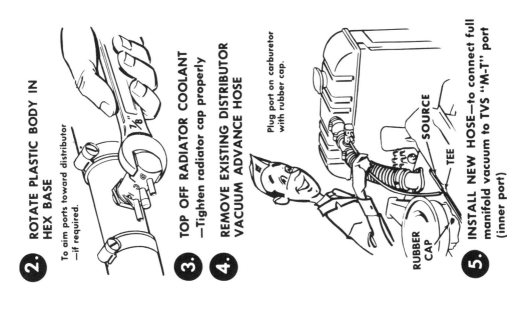

2. ROTATE PLASTIC BODY IN HEX BASE

To aim ports toward distributor —if required.

7/8"

3. TOP OFF RADIATOR COOLANT —Tighten radiator cap properly

4. REMOVE EXISTING DISTRIBUTOR VACUUM ADVANCE HOSE

Plug port on carburetor with rubber cap.

SOURCE

TEE

RUBBER CAP

5. INSTALL NEW HOSE—to connect full manifold vacuum to TVS "M-T" port (inner port)

Tap into source for full manifold vacuum at idle, using Tee or DEALER supplied fitting as required.

IMPORTANT — To accomplish the best possible result from the installation of a GM Exhaust Emission Control Kit, certain basic items must first be checked on each vehicle.

A — Determine that there is a minimum of ¾" clearance between the exhaust system and any point along the underbody. Adjust if necessary. The installation of this kit can cause increased exhaust temperatures, and proper clearance will reduce the possibility of excessive heat build up in the floor pan.

B — Check ignition system to assure that all spark plugs are firing properly.
   — Reconnect any loose ignition wires.
   — Check for crossed ignition wiring.

1. INSTALL TVS

—At end of hose

Reposition hose or cut 2" off to allow for extra length of TVS. Rotate sleeve so TVS clears engine accessories.

2¼" RADIATOR HOSE

DRAIN COOLANT AS REQUIRED AND SAVE FOR REFILL

OR

—Install in straight section.
Cut 1" out of center of hose.

ROTATE SLEEVE SO TVS CLEARS ENGINE ACCESSORIES.

GM Device—Cont'd

6. INSTALL NEW HOSE—to connect distributor to TVS "D" port (outer port)

TO MANIFOLD VACUUM

Leave plastic filter cap on TVS "C" port (center port)

Then—

7. SET IGNITION TIMING TO MANUFACTURER'S SPECIFICATIONS
—Follow manufacturer's latest service information
—DISCONNECT VACUUM ADVANCE HOSE TO DISTRIBUTOR AND PLUG
—If carburetor includes a HOT IDLE COMPENSATOR hold closed (or cover vent) during timing, idle speed, and mixture adjustment

8. IF AVAILABLE, SET UP EXHAUST GAS ANALYZER ON CAR
—Zero adjust analyzer on fresh air before inserting sample line in exhaust pipe

9. ADJUST IDLE SPEED (use accurate tachometer)
—600 RPM for AUTO. transmission (in DRIVE)
—700 RPM for MANUAL transmission (in NEUTRAL)
—engine at NORMAL OPERATING TEMPERATURE
—air conditioning OFF

10. ADJUST IDLE MIXTURE
—TO 1.5% CO or 14:1 A/F (Air fuel ratio) using EXHAUST GAS ANALYZER

ALTERNATE METHOD FOR SETTING IDLE MIXTURE

If analyzer is not available proceed as follows:
FIRST, set idle to speed shown in line (A) below. SECOND, adjust mixture screws (BALANCED) for BEST IDLE. THIRD, turn mixture screws LEANER (turn equally clockwise) until idle drops to speed shown in line (B) below.

	AUTO. TRANS. IN DRIVE 620—1 BBL CARB 640—2 & 4 BBL	MANUAL TRANS. IN NEUTRAL
(A) IDLE SPEED RPM FOR BEST IDLE	600	740
(B) LEAN MIXTURE SCREWS UNTIL SPEED DROPS TO		700

If carburetor has a separate adjustment for cold FAST IDLE, reset to manufacturer's specification.

11. RECHECK SETTINGS—
—If air cleaner is off to adjust idle speed and mixture RECHECK settings with air cleaner installed and READJUST as necessary

12. SEAL IDLE SPEED & MIXTURE SCREWS
—Remove dirt and film (use Solvent if required)
—Apply sealer to cover screw and spring
—Form "DO NOT TURN" label in "U" shape and apply over screw head and sealer.

DO NOT

BRIGHT GREEN ADHESIVE SEALER

13. RECONNECT vacuum advance hose to distributor

14. INSTALL GREEN STICKER
—In a conspicuous location underhood

15. INSTALL WINDOW STICKER
—Sign name and date then apply to REAR WINDOW.
(if regulations prohibit then apply to SIDE WINDOW)
YOU TOO MAY TAKE CREDIT FOR YOUR CONTRIBUTION TO CLEANER AIR THROUGH A GOOD JOB. REMIND THE OWNER TO HAVE THE EMISSION CONTROL ADJUSTMENTS CHECKED ANNUALLY.

GM Device—Cont'd

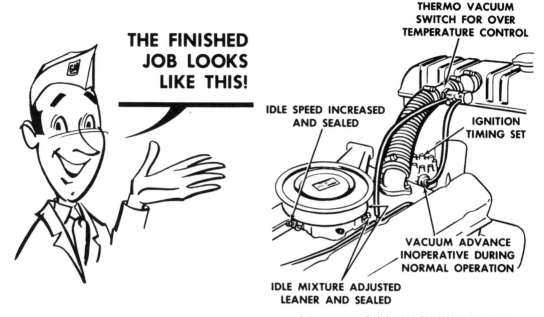

THE FINISHED JOB LOOKS LIKE THIS!

THERMO VACUUM SWITCH FOR OVER TEMPERATURE CONTROL

IDLE SPEED INCREASED AND SEALED

IGNITION TIMING SET

VACUUM ADVANCE INOPERATIVE DURING NORMAL OPERATION

IDLE MIXTURE ADJUSTED LEANER AND SEALED

It is recommended that the EMISSION CONTROL ADJUSTMENTS be checked ANNUALLY

Pure Power Device

PurePower
HIGH PERFORMANCE
IGNITION & EMISSION CONTROL SYSTEMS

AIR QUALITY PRODUCTS, INC.
950 NORTH MAIN STREET   ORANGE, CALIFORNIA 92667

Pure Power is accredited by California's Air Resources Board for installation on used motor vehicles under 6001 lbs. gross vehicle weight. This system is not to be installed on four cylinder engines, engines less than 140 cubic inch displacement, or having distributors without centrifical advance (Ford Loadamatic).

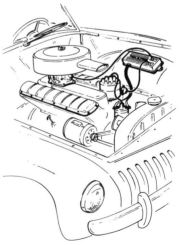

The Pure Power unit mounts on the fender panel, and both vacuum and electric connections are made to the engine

Pure Power Device—Cont'd

All Pure Power controls, both vacuum and electric, are inside this plastic box. Installation is simple

PRE-INSTALLATION STEPS

A HOOK UP ENGINE ANALYZER — including combustion analyzer — an HC/CO measurement instrument is preferred — if available.

B SET IGNITION TIMING & DWELL. Check manufacturer's timing specifications. Accurate timing adjustment is necessary for proper PURE POWER operation and effective emission control.

C CHECK FOR ENGINE MISFIRES. Check for misfire at idle and again at 2500 RPM — engine misfires should be corrected now to insure best PURE-POWER operation and effective emission control even though later installation of PURE POWER may correct misfires, because of its built-in high-voltage electronic ignition system.

D CHECK (PCV) CRANKCASE VENTILATION SYSTEM

E CHECK CARBURETOR & AIR FILTER. Carburetor should be checked at idle for manufacturer's specified RPM and mixture adjustment. At 2,500 RPM check for excessive rich condition (AFR shows slight increase over idle setting). Correct carburation is necessary for best emission control.

INSTALLATION PREPARATION

PURE-POWER must be set up BEFORE INSTALLATION to operate properly with the particular engine on which you are installing it.

NUMBER OF CYLINDERS Adjustments are made for an 8 cylinder engine by cutting the jumper wire on bottom of unit. If engine is 6 cylinder — DO NOT CUT JUMPER. PURE POWER is not designed for operation on four cylinder vehicles.

ENGINE TIMING Adjustments are made for engine timing by inserting one of five timing control modules in the socket on the bottom of the unit right next to the 6 - 8 cylinder jumper wire. Select the correct module color from the adjustment table, then insert it carefully into the two circuit board holes, seating it firmly.

1 CYLINDER SELECTION Check engine for number of cylinders — cut jumper if 8 cyl DO NOT CUT IF 6 CYLINDER

2 SPECIFICATIONS CHECK Check engine timing specifications and select correct control module from adjustment table

TIMING CONTROL MODULE

VEHICLE TIMING °BTDC	CONTROL MODULE COLOR
0 - 1	RED
2 - 3	RED
4 - 5	YELLOW
6 - 7	BLUE
8 - 9	GREY
10 – UP	BLACK

SELECTION TABLE

3 TIMING ADJUSTMENT Install timing control module in socket on circuit board

4 SEAL UNIT Fill out information blanks on seal and seal control module access hole

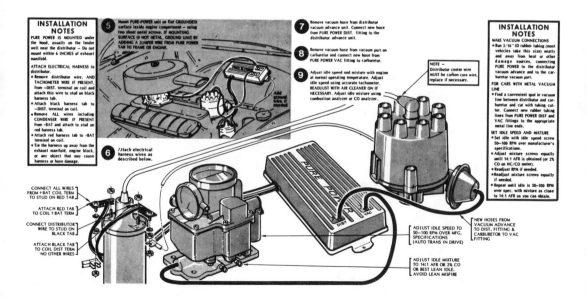

INSTALLATION NOTES

PURE POWER IS MOUNTED under the hood, usually on the fender well near the distributor — Do not mount within 6 INCHES of exhaust manifold.

ATTACH ELECTRICAL HARNESS to distributor.
• Remove distributor wire, AND TACHOMETER WIRE IF PRESENT, from –DIST. terminal on coil and attach this wire to stud on black harness tab.
• Attach black harness tab to –DIST. terminal on coil.
• Remove ALL wires including CONDENSER WIRE IF PRESENT from +BAT and attach to stud on red harness tab.
• Attach red harness tab to +BAT terminal on coil.
• Tie the harness up away from the exhaust manifold, engine block, or any object that may cause harness or hose damage.

5 Mount PURE-POWER unit on flat GROUNDED surface inside engine compartment — using two sheet metal screws. IF MOUNTING SURFACE IS NOT METAL, GROUND UNIT BY ADDING A JUMPER WIRE FROM PURE POWER TAB TO FRAME OR ENGINE.

Add ground wire if needed

6 Attach electrical harness wires as described below.

7 Remove vacuum hose from distributor vacuum advance unit. Connect new hose from PURE POWER DIST. fitting to the distributor advance unit.

8 Remove vacuum hose from vacuum port on carburetor and connect new hose from PURE POWER VAC fitting to carburetor.

9 Adjust idle speed and mixture with engine at normal operating temperature. Adjust idle speed using accurate tachometer. READJUST WITH AIR CLEANER ON IF NECESSARY. Adjust idle mixture using combustion analyzer or CO analyzer.

NOTE — Distributor center wire MUST be carbon core wire, replace if necessary.

INSTALLATION NOTES

MAKE VACUUM CONNECTIONS
• Run 3/16" ID rubber tubing (most vehicles take this size) neatly and away from heat or other damage sources, connecting PURE POWER to the distributor vacuum advance and to the carburetor vacuum port.

FOR CARS WITH METAL VACUUM LINE
• Find a convenient spot in vacuum line between distributor and carburetor and cut with tubing cutter. Connect new rubber tubing lines from PURE POWER DIST and VAC fittings to the appropriate metal line ends.

SET IDLE SPEED AND MIXTURE
• Set idle with idle speed screw 50–100 RPM over manufacturer's specifications.
• Adjust mixture screws equally until 14:1 AFR is obtained (or 2% CO on HC/CO meter).
• Readjust RPM if needed.
• Readjust mixture screws equally if needed.
• Repeat until idle is 50–100 RPM over spec. with mixture as close to 14:1 AFR as you can obtain.

CONNECT ALL WIRES FROM +BAT COIL TERM TO STUD ON RED TAB

ATTACH RED TAB TO COIL +BAT TERM

CONNECT DISTRIBUTOR WIRE TO STUD ON BLACK TAB

ATTACH BLACK TAB TO COIL DIST TERM NO OTHER WIRES

ADJUST IDLE SPEED TO 50–100 RPM OVER MFG. SPECIFICATIONS (AUTO TRANS IN DRIVE)

ADJUST IDLE MIXTURE TO 14:1 AFR OR 2% CO OR BEST EVEN IDLE. AVOID LEAN MISFIRE

NEW HOSES FROM VACUUM ADVANCE TO DIST. FITTING & CARBURETOR TO VAC FITTING

Pure Power Device—Cont'd

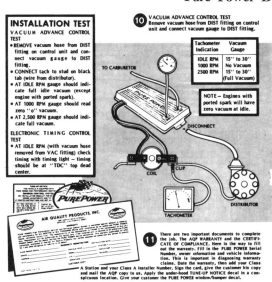

INSTALLATION TEST

VACUUM ADVANCE CONTROL TEST

- REMOVE vacuum hose from DIST fitting on control unit and connect vacuum gauge to DIST fitting.
- CONNECT tach to stud on black tab (wire from distributor).
- AT IDLE RPM gauge should indicate full idle vacuum (except engine with ported spark).
- AT 1000 RPM gauge should read zero "o" vacuum.
- AT 2,500 RPM gauge should indicate full vacuum.

ELECTRONIC TIMING CONTROL TEST

- AT IDLE RPM (with vacuum hose removed from VAC fitting) check timing with timing light – timing should be at "TDC" top dead center.

10 VACUUM ADVANCE CONTROL TEST
Remove vacuum hose from DIST fitting on control unit and connect vacuum gauge to DIST fitting.

Tachometer Indication	Vacuum Gauge
IDLE RPM	15" to 30"
1000 RPM	No Vacuum
2500 RPM	15" to 30" (Full Vacuum)

NOTE – Engines with ported spark will have zero vacuum at idle.

11 There are two important documents to complete the job. The AQF WARRANTY and the CERTIFICATE OF COMPLIANCE. Here is the way to fill out the warranty. Fill in the PURE POWER Serial Number, owner information and vehicle information. This is important in diagnosing warranty claims. Date the warranty, then add your Class A Station and your Class A Installer Number. Sign the card, give the customer his copy and mail the AQP copy to us. Apply the under-hood TUNE-UP decal in a conspicuous location. Give your customer the PURE POWER window/bumper decal.

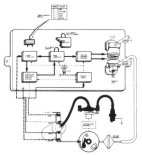

THEORY OF OPERATION

The PURE POWER unit is part of an emission control system used to bring about significant reduction in three harmful automotive exhaust byproducts; carbon monoxide, unburned hydrocarbons and oxides of nitrogen.

In the block diagram, functions of PURE POWER components are shown. The Timing Computer is set up during installation preparation by inserting a Timing Control Module. The TCM is used to provide an amount of electronic ignition delay equal to the advance originally built in by the vehicle manufacturer. This gives PURE POWER full control of ignition timing, a key factor in reducing emissions.

The Mode Computer alters the action of the Timing Computer to compensate for differences in 6 and 8 cylinder engines. Since PURE POWER operates on signals from the distributor breaker points the number of breaker point closures per engine revolution varies between 6 and 8 cylinder engines.

The Vacuum Advance Control section is programmed to energize the Vacuum Cut-off Valve as the engine speeds up from idle to 900 RPM. At about 1800 RPM the valve reopens. The sequence then is full vacuum advance at idle, cut-off between 900 and 1800 RPM, and full vacuum advance above. This valving tends to eliminate overheating during prolonged idle periods, reduce emissions during dirty acceleration states and restore full timing above 35 mph to maintain acceptable fuel economy and drive quality.

Actual firing of the cylinder spark plugs is controlled by the Timing Computer via the built in Capacitive Discharge Ignition System. Firing of the C/D system is retarded at idle, with respect to "normal" timing by action of the plug-in Timing Control Module. The electronic delay is gradually released during acceleration and the programmed cut-out of vacuum advance, at 900 to 1800 RPM, modifies the total vehicle advance curve to reduce emissions. During deceleration, cut-off occurs at 1300 RPM and vacuum advance returns at 700 - 800 RPM.

STP POLLUTION CONTROL DEVICE

INSTALLATION INSTRUCTIONS

CLASS "AA"–"A" VEHICLES
(50-140 CUBIC INCH DISPLACEMENT)

LIQUID COOLED VEHICLES
INSTALLATION INSTRUCTIONS CLASS "AA"–"A"
(50 through 140 Cubic Inch Displacement)

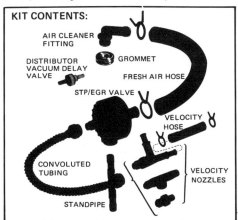

KIT CONTENTS:
- AIR CLEANER FITTING
- DISTRIBUTOR VACUUM DELAY VALVE
- GROMMET
- FRESH AIR HOSE
- STP/EGR VALVE
- VELOCITY HOSE
- CONVOLUTED TUBING
- VELOCITY NOZZLES
- STANDPIPE

STEP 1 – INSTALL THE VALVE

ALTERNATE INSTALLATION

PREFERRED INSTALLATION

VELOCITY NOZZLE

EGR VALVE

VACUUM DELAY VALVE

Use the STP/EGR Valve and Standpipe Assembly to determine the best position and angle to mount the Valve Assembly keeping clear of throttle linkage, fuel and vacuum lines, etc.

CORRECT WAY TO INSTALL EGR VALVE

WRONG WAY

Once the proper Valve position is determined, drill a 5/16" pilot hole in the Heat Riser of the Intake Manifold just below the carburetor. If the engine does not have an Exhaust Heat Riser for preheating the fuel mixture, drill the hole into the Exhaust Manifold at an angle that will prevent direct flow of exhaust gases into the EGR Valve (see illustration).

Enlarge pilot hole to a 7/16" hole, keeping the proper angle with the drill. Tap the hole, using a 1/4 - 18 tapered pipe tap.

Once the hole is tapped, use a magnet to remove external chips and shavings. Start engine with part throttle, and blow any remaining shavings from hole.

Screw threaded Standpipe and EGR Valve Assembly into tapped hole. Face threaded Valve outlet fitting toward the direction of the carburetor connection. Install EGR Valve by hand, tighten with wrench at base. Do not insert tool into Valve to tighten, or damage may result.

STP Pollution Control Device—Cont'd

STEP 2— INSTALL VELOCITY NOZZLE

A - FOR VEHICLES WITH PCV SYSTEMS

Remove the PCV hose from its carburetor or manifold connection. Install the end of the "Y"-shaped Velocity Nozzle, opposite its threaded inlet, to the carburetor connection; use the short Velocity Nozzle Hose and Clamps supplied. Reinstall PCV Hose on side leg of Velocity Nozzle.

IMPORTANT: DO NOT REVERSE THESE HOSE CONNECTIONS

B - FOR VEHICLES WITHOUT PCV SYSTEMS

Select a location in the Intake Manifold directly below the carburetor for installation of Manifold Fitting (Part No. 534-17). Drill and tap a 1/4" pilot hole and enlarge to proper size with an "R" Drill. (.3390"). Tap the hole using a 1/8 - 27 tapered pipe tap. Use a magnet to remove chips and shavings. Install Manifold fitting and tighten securely.

C - FOR VEHICLES WITH PCV VALVE SCREWED DIRECTLY INTO INTAKE MANIFOLD

Remove PCV Valve from Manifold and install Manifold Fitting (Part No. 534-24). Reinstall PCV Valve into side leg of Manifold Fitting.

In all cases, install flexible Convoluted Tubing from EGR Valve outlet onto Velocity Nozzle or Manifold Fitting, as the case may be, and tighten nuts securely.

STEP 3— INSTALL VACUUM DELAY VALVE

The vacuum Delay Valve assembly is required only on vehicles with combination Vacuum and Centrifugal Distributor advance mechanisms. It is not required for vehicles with Vacuum-only or Centrifugal-only spark advance Distributors (see STP Application Chart).

When required, install distributor Vacuum Delay Valve between Distributor Diaphragm and Carburetor with blue side toward carburetor. DO NOT REVERSE: VALVE ONLY OPERATES PROPERLY WHEN INSTALLED AS PRESCRIBED. On dual diaphragm Distributors, install the Vacuum Delay Valve into the line leading to the Advance Diaphragm. (See illustration.)

STEP 4— INSTALL AIR CLEANER HOSE

Remove Air Cleaner and drill pilot hole with 3/8" sheet metal drill, in location where the supplied Air Cleaner Fitting can be installed with direct access to STP/EGR Valve fresh air inlet hose. (Install Fitting on clean air side of Filter Element).

Use a 7/8" chassis punch to enlarge 3/8" pilot hole and install supplied Rubber Grommet and Air Cleaner Fitting.

Install supplied 5/8" rubber hose from STP/EGR Valve fresh air inlet nipple to Air Cleaner Fitting, using supplied Hose Clamps.

TYPICAL 4 CYLINDER INSTALLATIONS

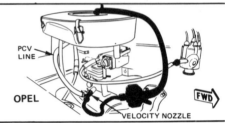

OPEL

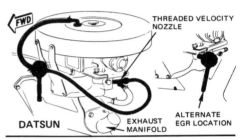

DATSUN

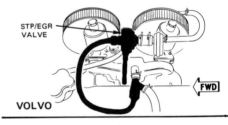

VOLVO

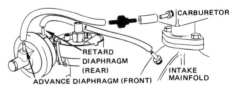

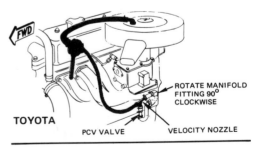

TOYOTA

GENERAL DIRECTIONS:

Readjust idle speed to car manufacturer's specifications, only if necessary. Install Maintenance Decal on Air Cleaner.

STP Pollution Control Device—Cont'd

VOLKSWAGEN ONLY
INSTALLATION INSTRUCTIONS

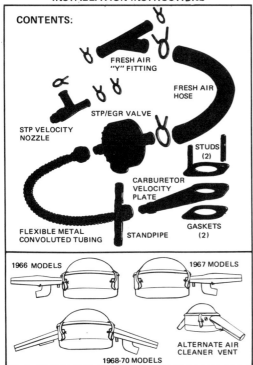

CONTENTS:

FRESH AIR "Y" FITTING

FRESH AIR HOSE

STP/EGR VALVE

STP VELOCITY NOZZLE

STUDS (2)

CARBURETOR VELOCITY PLATE

FLEXIBLE METAL CONVOLUTED TUBING

STANDPIPE

GASKETS (2)

1966 MODELS

1967 MODELS

1968-70 MODELS

ALTERNATE AIR CLEANER VENT

ALL MODELS EXCEPT AIR CONDITIONED :

LEFT SIDE PRE-HEATER PIPE

1. Car in neutral gear, hand brake on, ignition off.
2. Remove distributor vacuum advance line for easy access to left (driver's) side of pre-heater pipe (exhaust cross-over).

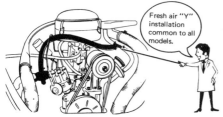

Fresh air "Y" installation common to all models.

3. Use EGR Valve and 3" Standpipe Assembly as a guage, and locate best position to mount Valve and Standpipe Assembly into left side of pre-heater pipe. Mark and centerpunch this location for drilling. Locate Valve Assembly so as to clear Heater and Air Cleaner Hoses. (See illustration for air conditioned models).
4. Drill 1/8" Hole, same angle as EGR Valve was located, into pre-heater pipe, then progressively enlarge with 1/4", 3/8" and finally 7/16" drills.
5. Tap 1/4 –18 tapered pipe thread into hole (bottom-out a standard tapered pipe tap). Start engine, rev several times to blow chips from tapped hole; stop engine.

6. Install EGR Valve and Standpipe Assembly into tapped hole, with fresh air inlet nipple up.
7. Install Fresh Air Inlet Hose
   A. Cut Crankcase Ventilation Hose 3" from Air Cleaner.
   B. Install Fresh Air "Y" into Ventilator Hose at cut, with 5/8" side leg pointing toward EGR Valve at left, or to the right on air conditioned models.
   C. Install 22" of 5/8" Fresh Air Hose onto EGR Valve and side leg of "Y" (10") on air conditioned models. Use supplied clamps at all new hose connection locations.

8. **FOR MANUAL TRANSMISSION MODELS ONLY:**

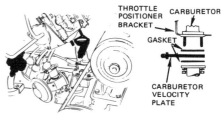

THROTTLE POSITIONER BRACKET

CARBURETOR

GASKET

CARBURETOR VELOCITY PLATE

A. Slip distributor cap off. Disconnect fuel line to carburetor at fuel pump.
B. Remove and save two washers and nuts from bottom of studs with 13mm wrench.
C. Remove and discard stock studs with small pliers (while carburetor is in place).
D. Slide carburetor and throttle positioner bracket to left and clean old gasket from manifold. Install supplied longer studs into carburetor base from bottom. Tighten with pliers.
E. Install furnished new carburetor gasket onto carburetor base from bottom, slip carburetor Velocity Plate onto carburetor base, add second gasket and slip carburetor with studs back into manifold holes.

DISTRIBUTOR VACUUM ADVANCE LINE

CARBURETOR VELOCITY PLATE

FUEL LINE

CONVOLUTED TUBING

F. Re-install washers and nuts, tighten.
G. Re-install fuel line and Distributor Cap.
H. Install flexible convoluted tubing onto bottom of EGR Valve and carburetor velocity plate. Tighten nuts with 17mm and 19mm wrenches.

9. **FOR AUTOMATIC TRANSMISSION MODELS ONLY:**

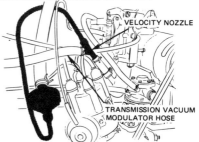

VELOCITY NOZZLE

TRANSMISSION VACUUM MODULATOR HOSE

A. Remove at manifold the vacuum line that goes from intake manifold to Transmission Vacuum - Modulator.
B. Install supplied Velocity Nozzle and short 1/2" hose (and clamps) supplied onto exposed 1/2" vacuum manifold fitting.
C. Shorten the existing Transmission Vacuum—Modulator hose as necessary (approximately 2") and reinstall vacuum hose onto side leg of Velocity Nozzle.
D. Install flexible convoluted tubing onto bottom of EGR Valve and onto threaded end of Velocity Nozzle. Tighten tubing nuts with 17mm and 19mm wrenches

STP Pollution Control Device—Cont'd

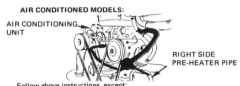

AIR CONDITIONED MODELS:

AIR CONDITIONING UNIT

RIGHT SIDE PRE-HEATER PIPE

Follow above instructions, except:
Use 4" Standpipe and install EGR Valve and Standpipe Assembly into right side of pre-heater pipe, as illustrated above.

ALL MODELS   10. *Start engine, adjust idle speed to manufacturer's specifications for individual vehicle year.*
11. *Install maintenance decal on Air Cleaner.*

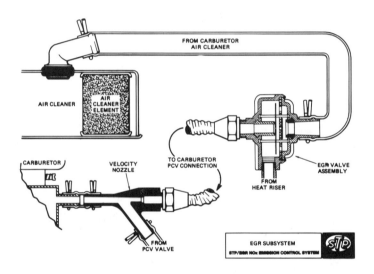

Typical connections for the STP exhaust gas recirculation valve

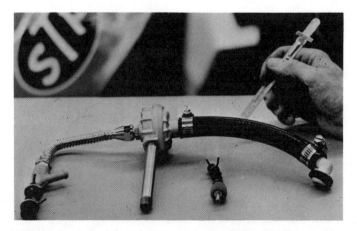

STP's device is mainly exhaust gas recirculation. Earlier versions used a spark bleed valve, but later ones use a delay valve

STP POLLUTION CONTROL DEVICE

INSTALLATION INSTRUCTIONS

CLASS "B" THROUGH "F" VEHICLES
(141 CUBIC INCH DISPLACEMENT AND OVER)

STEP 1 INSTALL THE EGR VALVE

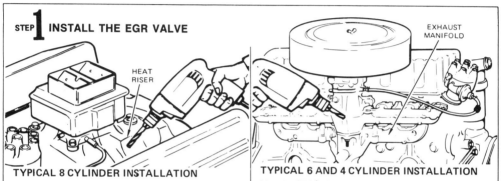

TYPICAL 8 CYLINDER INSTALLATION

Use the STP/EGR Valve and Standpipe assembly to determine the best position and angle to mount the Valve Assembly, keeping clear of throttle linkage, etc.

Once the proper position is determined, drill a 5/16" pilot hole in the Heat Riser Portion of the Intake Manifold. DO NOT INSTALL EGR VALVE DIRECTLY INTO EXHAUST MANIFOLD. Enlarge to a 7/16" hole, keeping the proper angle with the drill.

Tap the hole using a 1/4—18 tapered pipe tap.

Once the hole is tapped, use a magnet to remove external chips and shavings. Start engine with part throttle, and blow remaining chips from hole.

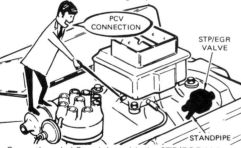

Screw threaded Standpipe with the STP/EGR Valve into tapped hole. Face Valve outlet fitting (threaded) toward the direction of the PCV Carburetor connection.

TYPICAL 6 AND 4 CYLINDER INSTALLATION

Use the STP/EGR Valve and Standpipe assembly to determine the best position and angle to mount the Valve Assembly, keeping clear of throttle linkage, etc.

Once the proper position is determined, drill a 5/16" pilot hole in the Heat Riser portion of the Intake Manifold. DO NOT INSTALL EGR VALVE DIRECTLY INTO EXHAUST MANIFOLD. Enlarge to a 7/16" hole, keeping the proper angle with the drill.

Tap the hole using a 1/4—18 tapered pipe tap.

Once the hole is tapped, use a magnet to remove external chips and shavings. Start engine with part throttle, and blow remaining chips from hole.

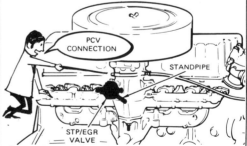

Screw threaded Standpipe with the STP/EGR Valve into tapped hole. Face Valve outlet fitting (threaded) toward the direction of the PCV Carburetor connection.

INSTALL EGR VALVE BY HAND, TIGHTEN WITH WRENCH AT BASE. DO NOT INSERT TOOL INTO VALVE TO TIGHTEN, OR DAMAGE MAY RESULT.

STP Pollution Control Device—Cont'd

STEP 2 HOOK UP TO VELOCITY NOZZLE AND RE-CONNECT PCV LINE

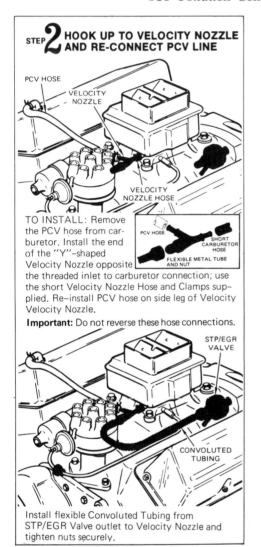

PCV HOSE

VELOCITY NOZZLE

VELOCITY NOZZLE HOSE

TO INSTALL: Remove the PCV hose from carburetor. Install the end of the "Y"-shaped Velocity Nozzle opposite the threaded inlet to carburetor connection; use the short Velocity Nozzle Hose and Clamps supplied. Re-install PCV hose on side leg of Velocity Velocity Nozzle.

PCV HOSE — SHORT CARBURETOR HOSE — FLEXIBLE METAL TUBE AND NUT

Important: Do not reverse these hose connections.

STP/EGR VALVE

CONVOLUTED TUBING

Install flexible Convoluted Tubing from STP/EGR Valve outlet to Velocity Nozzle and tighten nuts securely.

STEP 3 CONNECT AIR CLEANER HOSE

AIR CLEANER (Install in bottom side where practical)

Remove Air Cleaner and drill pilot hole with 3/8" sheet metal drill, in location where the supplied Air Cleaner Fitting can be installed with direct access to STP/EGR Valve fresh air inlet hose. (Install Fitting on clean air side of Filter Element).

AIR CLEANER FITTING

RUBBER GROMMET

Use a 7/8" chassis punch to enlarge 3/8" pilot hole and install supplied Rubber Grommet and Air Cleaner Fitting.

5/8" RUBBER HOSE

Install supplied 5/8" rubber hose from STP/EGR Valve fresh air inlet nipple to Air Cleaner Fitting, using supplied Hose Clamps.

INSPECTION AND MAINTENANCE

The STP Pollution Control Device is an Exhaust Gas Recirculation (EGR) system designed to provide years of trouble-free service. It consists of the EGR Valve and its plumbing, along with a vacuum-advance Delay Valve* for vehicles with combination vacuum and centrifugal advance distributors. The following inspection procedures should be followed every 12,000 miles of operation:

1. Inspect Velocity Nozzle (or VW carburetor plate) for internal carbon-deposit buildup. Remove any carbon with stiff bottle brush; if orifice cannot be cleared, replace part.
2. Check flexible metal tube and hose for leaks; replace any worn components.
3. Check EGR Valve function with vehicle in neutral idle; flexible tube should be hot. Replace Valve if flex tube fails to become heated within two minutes.
4. If vehicle is equipped with an STP Vacuum Delay Valve* proceed as follows:
   A. Connect vacuum guage into distributor vacuum line between STP Delay Valve* and distributor diaphragm.
   B. With engine running in neutral at 2000 RPM, check to see if at least 15" of vacuum is restored to distributor diaphragm within one minute.

C. Delay Valve* cannot be cleaned or serviced, and is normally replaced no oftener than every 12,000 miles.

VEHICLE TUNE-UP PROCEDURE

Disconnect STP Device as follows:
1. Remove Vacuum Delay Valve* from distributor vacuum line, if so equipped.
2. Disconnect fresh air hose between EGR Valve and air cleaner.
3. Disconnect metal flex line and EGR Nozzle assembly from base of carburetor; remove PCV hose from side leg of Nozzle and connect to carburetor.
   A. VOLKSWAGEN ONLY: Disconnect metal flex line from carburetor plate. Plug hole in plate with suitable plug to prevent air leak.
4. Tune to manufacturer's specifications.
5. Reconnect STP Device and reinstall Vacuum Delay Valve* in vacuum line.

*NOTE: Vacuum Delay Valve is not used on Volkswagens or any other vehicle with vacuum-only or centrifugal-only distributor advance systems.

STP Pollution Control Device—Cont'd

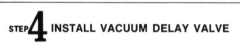

STEP 4 **INSTALL VACUUM DELAY VALVE**

Install Distributor Vacuum Delay Valve between Distributor Diaphragm and Carburetor.

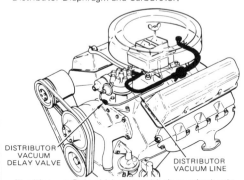

DISTRIBUTOR
VACUUM
DELAY VALVE

DISTRIBUTOR
VACUUM LINE

The Vacuum Delay Valve assembly is required only on vehicles with combination Vacuum and Centrifugal Distributor advance mechanisms. It is not required for vehicles with Vacuum-only or Centrifugal-only spark advance Distributors (see STP Application Chart).

When required, install distributor Vacuum Delay Valve between Distributor Diaphragm and Carburetor with blue side toward carburetor. DO NOT REVERSE: VALVE ONLY OPERATES PROPERLY WHEN INSTALLED AS PRESCRIBED. On dual diaphragm Distributors, install the Vacuum Delay Valve into the line leading to the Advance Diaphragm (see illustration below):

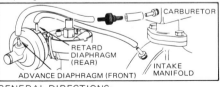

CARBURETOR

RETARD
DIAPHRAGM
(REAR)

INTAKE
MANIFOLD

ADVANCE DIAPHRAGM (FRONT)

GENERAL DIRECTIONS:
Readjust idle speed to car manufacturer's specifications, only if necessary.
Install Maintenance Decal on Air Cleaner.

© 1973 STP CORPORATION

For further information, contact
STP CORPORATION
WEST COAST OPERATIONS
929 Olympic Boulevard
Santa Monica, Calif. 90404
(213) 394-2771

KIT CONTENTS

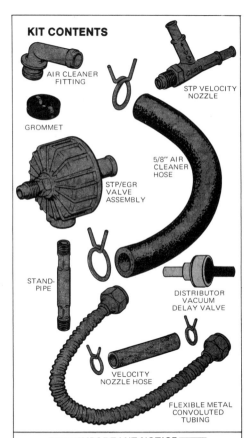

AIR CLEANER
FITTING

STP VELOCITY
NOZZLE

GROMMET

5/8" AIR
CLEANER
HOSE

STP/EGR
VALVE
ASSEMBLY

STAND-
PIPE

DISTRIBUTOR
VACUUM
DELAY VALVE

VELOCITY
NOZZLE HOSE

FLEXIBLE METAL
CONVOLUTED
TUBING

■■■IMPORTANT NOTICE■■■
Refer to STP Pollution Control Device Application Chart for proper selection of EGR Valve Assembly and Distributor Vacuum Delay Valve.

EGR VALVE LETTER CODE	
AA – 50 to 89 CID	D – 251 to 300 CID
A – 90 to 140 CID	E – 301 to 375 CID
B – 141 to 200 CID	F – 376 and over CID
C – 201 to 250 CID	

Echlin Device

Installation Instructions

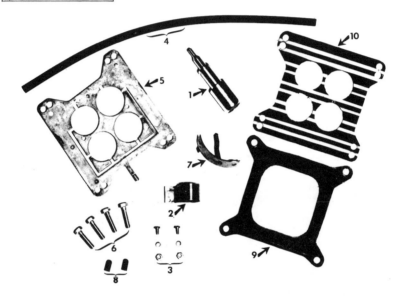

• 1 Energizer   • 2 Bracket, energizer mounting • 3 Screws, nuts, and lock-washers — bracket mounting — 10-32 x ½ • 4 Hose, connecting • 5 Plate, wave-guide adapter • 6 Cap Screws* • 7 Bracket, distributor vacuum advance lockout • 8 Cap, vacuum port • 9 Gasket, plate to manifold • 10 Gasket, carburetor to plate
Some kits contain studs.

ENGINE CONDITION IS IMPORTANT!

Although it is not required by the provisions of the law, it is recommended that a complete diagnosis of the engine be performed to determine its condition prior to the installation of the emissions control system.

An engine with bad rings or bad valves or worn ignition system components, i.e., points, condenser, cap, rotor, spark plug wires, or spark plugs can be emitting at higher than desired levels even though the control devices are operating. Fuel system (carburetor) malfunction can also be a major contributor to higher than normal emissions.

The diagnosis of vehicle condition should include an ignition and electrical system check, a cylinder leakage test, and/or a compression test. Any deficiencies uncovered should be brought to the attention of the vehicle owner. He should be informed that the effectiveness of emissions control, satisfactory fuel economy, driveability, and safety are dependent upon the basic condition of the vehicle.

The vehicle should then be rechecked periodically at normal service intervals to maintain operating efficiency as well as to assure low level of exhaust emissions.

Echlin Device—Cont'd

ENGINE PREPARATION

1. Remove the air cleaner and carburetor from the engine.
2. Check the carburetor for correct application, damage, malfunction, and leaks.
3. Remove the carburetor mounting studs from manifold (use stud remover like NS4916).
4. Clean the manifold gasket surfaces and heat passages.
5. Check the gasket surface for flatness using a small straight edge.

INSTALLATION

6. If studs are supplied in kit install in the manifold at this time.
7. Assemble hose to the waveguide adapter plate nipple.
8. Place the orange striped gasket on the manifold.
9. Install the plate with waveguide channels facing DOWN toward the intake manifold.
10. Install the white striped gasket on top of the plate and install the carburetor.
11. Adjust the choke rod and check all the external carburetor adjustments including bowl vent, unloader, choke pulloff, and all linkage.
12. If so equipped, remove idle mixture screw limiter cap. Seat screw lightly and back out 1½ to 2 turns.
    Note: Some carburetors use a hidden allen screw to limit the adjustment range of limiter screws. Check the appropriate shop manual for removal procedure.
13. Remove the vacuum advance hose from carburetor.
14. Seal carburetor vacuum port with cap provided.
15. Remove the vacuum advance unit from the distributor. Replace with the indexing bracket provided.
16. Place the air cleaner on the carburetor in the proper position. Determine the location and hose routing for the sonic energizer.
17. Mark mounting hole location on air cleaner so that the top of the energizer will be ½″ or more below the top of the air cleaner.

The main parts of the Echlin device are the carburetor plate and the energizer, which is a special kind of air bleed

18. Remove the air cleaner from the carburetor and drill the holes for the mounting screws.
19. Insert the energizer in the bracket and assemble to the air cleaner using the screws, nuts, and lockwashers provided.
20. Place the air cleaner on the carburetor in the correct position, position the connector hose, cut to length allowing 1 - 2 inches extra length and install onto energizer.

ADJUSTMENT & TUNING

21. Start the engine and check for fuel and vacuum leaks.
22. Allow engine to warm up at a fast idle.
23. Check and adjust dwell to manufacturers' specification.
24. Set timing (with engine operating at hot idle) at 2° B.T.D.C.
25. Adjust idle RPM to manufacturers' specifications.
26. Adjust idle mixture to 2% CO/14.0:1 air/fuel ratio. Readjust idle speed as necessary while maintaining the 2% CO setting.
27. Apply Locktite "290" (green) to idle mixture screw threads to "lock" adjustment.

Echlin Device—Cont'd

28. Apply service information decal. This decal must be prominently placed on top of the air cleaner or fan shroud. Be sure to clean the surface, being certain that the area is free of oil, dirt, etc. Pull ½ of the backing off starting at the center — carefully position the decal on the cleaned area avoiding wrinkles, then slowly re- move remaining backing.

29. Place the ¾ x 2″ decal on energizer bracket to identify the unit.

30. Be certain to give the Owner's Informa- tion Manual, included in the energizer kit, to the driver of the vehicle when it is delivered.

THE ECHLIN DEVICE is applicable to all 1966-70 vehicles except the following:

1. All Imported cars.
2. All Domestic vehicles with engines under 200 cubic inch displacement.
3. All engines equipped with multiple carburetors.
4. All engines equipped with fuel injection.
5. All engines equipped with full vacuum advance distributors.
6. All International Harvester Vehicles.

ATTENTION!

This car is equipped with THE ECHLIN DEVICE to control oxides of nitrogen emissions as required by the California Health & Safety Code and as approved by the California Air Resources Board.

The following engine modifications have been made and these specifications must be adhered to when servicing the engine:

1. Vacuum advance disconnected.
2. Carburetor adjusted to 2% C.O. (and/or 14.0 to 1 A.F.R.).
3. Timing spec. 2° B.T.C.
4. Wash air filter in the sonic generator in parts cleaning solvent or equivalent at carburetor air filter replacement intervals as specified in Owner's Manual.

THE ECHLIN MANUFACTURING COMPANY • Branford, Connecticut 06405

ATTENTION INSTALLERS

This decal must be prominently placed on the top of the air cleaner or fan shroud

Be sure to clean the surface, being certain that the area is free of oil, dirt etc.

Pull about 1/2 of this back off—starting at the center. Carefully position the decal on the cleaned area, avoiding wrinkles—then, slowly remove remaining back up. Press down firmly on the decal to completely seal the edges.

Be certain to give owner's information manual included in the energizer kit to the driver of the vehicle when it is delivered.

Thank you for cooperating in improving the quality of the air we breathe

ECHLIN MANUFACTURING COMPANY • Branford, Connecticut 06405

Perfect Circle (Retronox™) Device

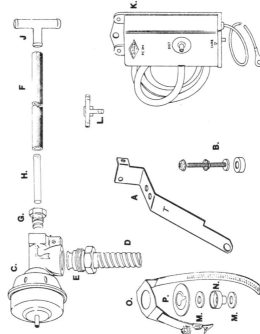

INSTALLATION INSTRUCTION SHEET
RETRONOX

A speed regulated system for control of exhaust emissions

Perfect Circle

A. EGR Valve Bracket STAMPED "T"
B. Bumper Assembly
C. EGR Valve
D. Corrugated Tube
E. Inverted Flare Nut
F. Green Silicone Hose
G. Compression Nut
H. ⅜ in. Hose Nipple
J. Metal Tee
K. Engine Speed Switch
L. Plastic Tee
M. Gaskets
N. Exhaust Inlet Cup
O. Exhaust Inlet Clamp
P. Tube Clamp Nut

ABBREVIATIONS
EGR—Exhaust Gas Recirculation. EGRV—Exhaust Gas Recirculation Valve.
PCV—Positive Crankcase Ventilation. ESS—Engine Speed Switch.

Perfect Circle (Retronox™) Device—Cont'd

MAINTENANCE INSTRUCTIONS — 12,000-MILE OR ANNUAL

1. At the EGR valve, disconnect line to exhaust pipe and line to PCV hose or intake manifold. Plug both lines. Detach EGRV from bracket but leave vacuum hose connected. Run engine at fast idle to retract valve plunger.

2. With plunger retracted, use small wire brush to clean deposits from inlet and outlet valve ports. Use bent stiff wire to clean valve seat and plunger.

3. Blow out debris from plunger end of valve only. DO NOT BLOW COMPRESSED AIR INTO VACUUM CONNECTION OR VENT HOLES.

4. DO NOT SOAK VALVE IN SOLVENT OR CARBURETOR CLEANER. DIAPHRAGM MAY BE DAMAGED AND VALVE WARRANTY VOIDED.

5. Stop engine and unplug lines. Clean out line to carburetor or manifold (including tee and all fittings). Mount EGRV on bracket and connect both lines. Read Part 1. of Step 7 and check system operation.

6. If vacuum hoses must be replaced, use same type material.

7. Tighten exhaust pipe clamp if loose. Replace gaskets if connection leaks.

8. Record maintenance below:

NOTE: There is no internal service for the engine speed switch.

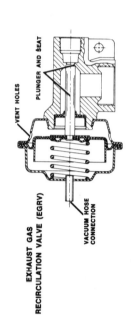

EXHAUST GAS RECIRCULATION VALVE (EGRV)

Labels: PLUNGER AND SEAT / VENT HOLES / VACUUM HOSE CONNECTION

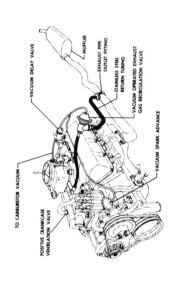

Labels: VACUUM DELAY VALVE / MUFFLER / EXHAUST PIPE OUTLET FITTING / STAINLESS STEEL RETURN TUBING / VACUUM OPERATED EXHAUST GAS RECIRCULATION VALVE / VACUUM SPARK ADVANCE / TO CARBURETOR VACUUM / POSITIVE CRANKCASE VENTILATION VALVE

INSTALLATION CHECK AND/OR TROUBLESHOOTING GUIDE

To check the operation of the system, run the engine in neutral above 1400 rpm (or at fast idle) and feel the EGR valve. It should become warm in 15-20 seconds and should continue to get hotter. If the valve and tubes do not get warm, look in the slotted holes around the EGR valve as the engine speed is changed from slow to fast and back to slow a few times. If the valve is functioning, it should be possible to see the black rubber diaphragm move back and forth past the holes.

1. If the diaphragm is moving, but the valve remains cool, the exhaust return line is plugged between the exhaust pipe and the valve or between the valve and the carburetor. Disassemble these parts and check the passages. See also that all connections are free of leaks.

2. If the diaphragm is not moving, it means that the vacuum is either not getting through or it is too low to operate the valve. Disconnect the hose at the carburetor and open the throttle enough to obtain a fast idle speed. There should be plenty of vacuum at the carburetor connection. If not, there is a leak somewhere in the existing equipment or the engine is in bad condition. Check PCV valve, engine compression, and all vacuum hoses.

3. If vacuum is available at the carburetor, reconnect the hose from the carburetor and disconnect the other hose that is attached to the plastic connector on the front of the switch. Run the engine above 1400 rpm again to see if the vacuum is getting through the switch. If it is not, check to see if: 1) the switch is mounted on a metal panel; 2) the mounting screw is tight and the lockwasher is under the head of the screw; 3) the lead wire goes to the battery positive (+) side of the ignition coil; and 4) the nut on the coil terminal post is tight. When all of these checks are OK, retest the switch to see if vacuum will pass through when the throttle is open. If it will not, the switch must be replaced.

4. If vacuum is reaching the speed switch, reconnect the hose from the carburetor and disconnect the other hose and perform the same check at the other end by disconnecting this hose from the speed switch. If vacuum is not getting through, the hose is either plugged, pinched, or reversed at the switch. The hose from the carburetor attaches to the metal connector on the bottom of the speed switch.

5. If the switch is all right, the next place to check is at the plastic tee. Disconnect the hose there that runs to the distributor (or the canister on 1970 vehicles) and see if vacuum reaches the tee. If the line to the tee is clear, place a finger over the open side of the tee and see if the EGR valve will now function when the engine is speeded up. If it still does not get warm, either the hose to it is plugged or the valve is defective.

6. If the valve works all right, but does not work when the hose from the distributor (or canister) is reconnected, then the vacuum spark advance actuator (or canister) is at fault.

NOTE: *When the EGR valve opens, the vacuum in the line will fall approximately 2 in. Hg. vacuum. As the engine drops about 200 rpm below cut-in speed, the EGR part of the system will cease to operate. The valve must not get hot when the engine is running below this speed.*

IMPORTANT! COMPLETE, SIGN, AND DATE OWNER'S WARRANTY. LEAVE WITH INSTALLATION INSTRUCTIONS AND STORE IN VEHICLE.

Perfect Circle (Retronox™) Device—Cont'd

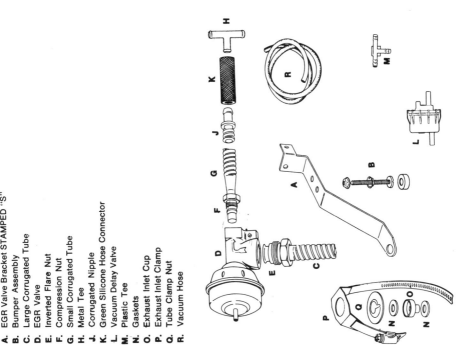

A. EGR Valve Bracket STAMPED "S"
B. Bumper Assembly
C. Large Corrugated Tube
D. EGR Valve
E. Inverted Flare Nut
F. Compression Nut
G. Small Corrugated Tube
H. Metal Tee
J. Corrugated Nipple
K. Green Silicone Hose Connector
L. Vacuum Delay Valve
M. Plastic Tee
N. Gaskets
O. Exhaust Inlet Cup
P. Exhaust Inlet Clamp
Q. Tube Clamp Nut
R. Vacuum Hose

ABBREVIATIONS
EGR—Exhaust Gas Recirculation. EGRV—Exhaust Gas Recirculation Valve
PCV—Positive Crankcase Ventilation. VDV—Vacuum Delay Valve.

INSTALLATION INSTRUCTION SHEET

RETRONOX

THE EXHAUST EMISSION CONTROL SYSTEM

Perfect Circle
DANA

Perfect Circle (Retronox™) Device—Cont'd

MAINTENANCE INSTRUCTIONS — 12,000-MILE OR ANNUAL

1. At the EGR valve, disconnect line to exhaust pipe and line to PCV hose or intake manifold. Plug both lines. Detach EGRV from bracket but leave vacuum hose connected. Run engine at fast idle to retract valve plunger. If EGR valve plunger does not retract in 15 to 30 seconds after opening throttle, replace vacuum delay valve.

2. With plunger retracted, use small wire brush to clean deposits from inlet and outlet valve ports. Use bent stiff wire to clean valve seat and plunger.

3. Blow out debris from plunger end of valve only. DO NOT BLOW COMPRESSED AIR INTO VACUUM CONNECTION OR VENT HOLES.

4. DO NOT SOAK VALVE IN SOLVENT OR CARBURETOR CLEANER. DIAPHRAGM MAY BE DAMAGED AND VALVE WARRANTY VOIDED.

5. Stop engine and unplug lines. Clean out line to carburetor or manifold (including tee and all fittings). Mount EGRV on bracket and connect both lines. Read Part 1. of Step 6 and check system operation.

6. If vacuum hoses must be replaced, use same type material.

7. Tighten exhaust pipe clamp if loose. Replace gaskets if connection leaks.

INSTALLATION ON EXHAUST PIPE

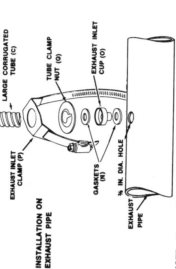

LARGE CORRUGATED TUBE (C)

TUBE CLAMP NUT (Q)

EXHAUST INLET CUP (O)

EXHAUST INLET CLAMP (P)

GASKETS (N)

⅜ IN. DIA. HOLE

EXHAUST PIPE

NOTE: All specifications MUST be met to assure a good seal. Both gaskets MUST be replaced if seal is broken.

ENGINE SPEED SWITCH & SOLENOID VACUUM VALVE

MUFFLER

EXHAUST PIPE OUTLET FITTING

STAINLESS STEEL RETURN TUBING

VACUUM OPERATED EXHAUST GAS RECIRCULATION VALVE

VACUUM SPARK ADVANCE

TO CARBURETOR VACUUM

POSITIVE CRANKCASE VENTILATION VALVE

INSTALLATION CHECK AND/OR TROUBLESHOOTING GUIDE.

1. To check the operation of the system, the engine must be warm and running in neutral at slow or hot idle speed (throttle must be closed). Shine a lite thru the slots in the EGR valve actuator section and watch the rubber diaphragm as you open the throttle to fast idle position and hold it there. Within about 5 seconds the valve should start to open, and after 15-30 seconds it should reach the end of its stroke. When you close the throttle, you should see the diaphragm respond immediately and quickly return to its closed position.

2. If the EGR valve opens suddenly and closes slowly, it means the vacuum delay valve is reversed—the hose from the carburetor should be connected to the green side of the Vacuum Delay Valve (VDV).

3. If the EGR valve opens but does not get hot, there is an obstruction somewhere in the line between the exhaust pipe and the EGR valve or between the EGR valve and the intake manifold.

4. If the diaphragm does not move, it could mean: a) One of the vacuum hoses just installed is plugged, pinched, or kinked—check all hoses. b) There is a leak in some existing vacuum operated system—check all tubes, hoses, connections, servos, diaphragms, and valves. c) The PCV valve is stuck open—replace it. d) The wrong nipple on the carburetor is being used—there should be no vacuum available unless the throttle is open. e) The VDV is inoperative—disconnect the vacuum line from the plastic tee and hold your finger over the end of the hose. Open the throttle. If there is no vacuum at your finger in about 30 seconds, the VDV should be replaced. Check operation of the EGR valve after each of the above steps is completed.

5. If trouble is still not found, there are two checks yet to be made. At the plastic tee, disconnect the hose which feeds the distributor. Plug the opening in the tee and open the throttle. If the EGR valve now functions properly, the fault is in the diaphragm of the vacuum spark advance actuator.

6. If the valve still does not function as specified, attach a vacuum hose to the EGR valve and run it directly to the carburetor nipple. When the throttle is opened, the valve should open immediately and close as soon as the throttle is closed. If this does not happen, replace the valve.

IMPORTANT! COMPLETE, SIGN, AND DATE OWNER'S WARRANTY. LEAVE WITH INSTALLATION INSTRUCTIONS AND STORE IN VEHICLE.

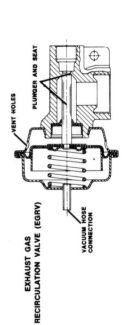

PLUNGER AND SEAT

VENT HOLES

VACUUM HOSE CONNECTION

EXHAUST GAS RECIRCULATION VALVE (EGRV)

Carter Device

INSTALLATION INSTRUCTIONS
CARTER EMISSION REDUCTION KIT

TYPICAL V-8 INSTALLATION

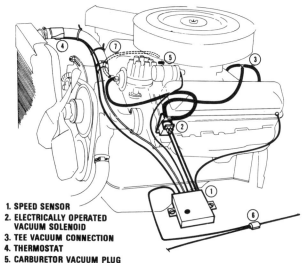

1. SPEED SENSOR
2. ELECTRICALLY OPERATED
   VACUUM SOLENOID
3. TEE VACUUM CONNECTION
4. THERMOSTAT
5. CARBURETOR VACUUM PLUG
6. SNAP WIRE SPLICE
7. EXISTING VACUUM HOSE (REMOVE)

STEP ONE
ATTACHING SPEED SENSOR

● Using Speed Sensor box as template, mark off the two mounting holes on an area of the **fender-well** in a convenient location making sure the length of the speed sensor wires will accommodate the chosen location.

● Punch two ⅛ inch diameter holes and mount Speed Sensor with two of the No. 10 Self-tapping screws furnished. **NOTE: SPEED SENSOR SHOULD BE MOUNTED WITH WIRES IN A DOWNWARD POSITION.** Attach **black wire** (ground) to convenient ground on engine, such as valve cover bolt, etc.

MOUNTING SENSOR

GROUND BLACK WIRE TO ENGINE

Carter Device—Cont'd

TYPICAL 6 CYLINDER INSTALLATION

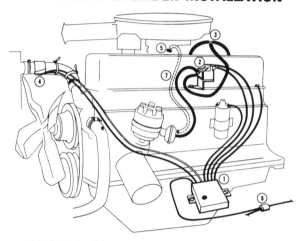

WARNING: DO NOT START INSTALLATION UNLESS IGNITION SWITCH IS "OFF." READ INSTRUCTIONS COMPLETELY BEFORE INSTALLATION.

STEP TWO
MOUNTING SOLENOID VALVE

● Locate manifold vacuum source. Generally the easiest source would be **carburetor choke break diaphragm hose, automatic transmission modulator hose,** or **air cleaner vacuum hose.** Mount solenoid valve to engine valve cover for easy hose hookup to distributor and manifold vacuum source, using existing bolt on valve cover. **NOTE: Make sure there is sufficient clearance for throttle lever and air cleaner.**
If valve cover is inaccessible, solenoid may also be mounted on fender-well using a punch and mounting with No. 10 sheet metal screw and washer furnished. **WARNING:** Do not pull off sponge rubber element from vacuum solenoid. This is a filter and should not be removed.

VALVE COVER

FENDER WELL

Carter Device—Cont'd

STEP THREE
CONNECTING DISTRIBUTOR TO VACUUM SOLENOID

● Remove existing line connecting distributor to carburetor and plug spark port of carburetor with cap supplied. For vehicles having a hard metal vacuum spark line, remove a section of the line leaving two inches extending from the distributor vacuum advance unit and two inches extending from the carburetor. Make sure to plug line extending from carburetor with cap supplied.

SOFT HOSE REMOVAL

HARD LINE REMOVAL

● The hose included in the kit is sufficient to connect the solenoid to the distributor and the solenoid to the manifold vacuum source. **Measure hose carefully before cutting.** Connect solenoid and distributor vacuum advance unit with new hose supplied. Connect hose to nipple on side of solenoid as indicated above.

STEP FOUR
CONNECTING MANIFOLD VACUUM SOURCE TO SOLENOID

● Connect other section of hose from solenoid to manifold vacuum source, cutting into existing vacuum hose using tee supplied. If hose is larger than tee, use clamps supplied to secure hose.

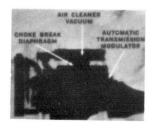

VACUUM SOURCES

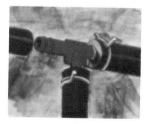

"T" TYPE CONNECTION

Carter Device—Cont'd

CONNECTING SPEED SENSOR WIRES

STEP FIVE
CONNECTING YELLOW SENSOR WIRES TO SOLENOID

- Attach **yellow lead wire** of speed sensor to the electrical tabs on the solenoid. **NOTE: Wires may be attached to either tab.**

STEP SIX

CONNECTING GREEN WIRE TO COIL

- Attach **green lead wire** to distributor terminal of coil (negative side of coil). Where push-on type terminals are used at the coil, cut terminal from green wire and use furnished splice connector to make connection with distributor wire.

STEP SEVEN
CONNECT RED POWER SOURCE WIRE

 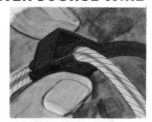

- Attach red wire to the following specified 12 volt power sources using splice connector supplied:

General Motors Corporation
Blue or yellow 18 gauge wire to windshield wiper or washer (1966-70)

Chrysler Corporation
Dark blue 16 gauge wire on ignition side of ballast resistor (1966-70)

Ford Motor Company
Red and green striped 18 gauge wire voltage regulator (1966-70)

American Motors Corporation
Yellow 18 gauge wire to voltage regulator (1966-69). Accessory connector at fuse box (1970)

Jeep Corporation
On 6 and 8 cylinder models, connect red wire to 12 volt lead at ignition ballast.

International Harvester Corporation
International Harvester 1966 through 1968 connect red wire with splice connector to #2 voltage regulator post #129 wire.

International Harvester 1969 and 1970 connect red wire with splice connector to #2 voltage regulator post #6 wire.

Make sure component wires, hoses, etc., do not interfere with throttle linkage or fan.

STEP EIGHT RESET ENGINE TIMING

- Distributor timing for all engine sizes, is to be reset at 4 degrees retard from vehicle manufacturers' standard timing specifications.

Carter Device—Cont'd

STEP NINE
SETTING SPEED SENSOR CONTROL FOR DISTRIBUTOR VACUUM ADVANCE

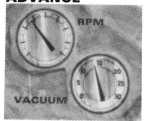

● Remove vacuum hose connection from Solenoid to Distributor **at Solenoid only.** Attach vacuum gauge to Solenoid. Start engine and bring up to R.P.M. specified for this vehicle by pulling back on throttle lever or having someone press on accelerator pedal. (Refer to chart for Specific Vehicle R.P.M.) Turn sensor screw SLOWLY clockwise until any vacuum reading is noted on gauge, at which time adjustment is completed. **NOTE:** To check setting, return engine to idle and then slowly open throttle until vacuum is noted on gauge. Engine R.P.M. should correspond with speed on chart. If not, repeat

REFER TO CHART FOR SPECIFIC VEHICLE R.P.M.

FORM 3530 PRINTED IN U.S.A. 3-73

adjustment. Replace hose from distributor to Solenoid. Clean speed sensor box with clean dry cloth. Apply decal to face of box so as to cover adjusting screw hole.

This decal must be installed to comply with California Emission Regulations.

STEP TEN
ADJUSTING IDLE MIXTURE

● Adjust idle mixture to lean-best idle and adjust speed to manufacturer's specifications. **LEAN-BEST IDLE:** Turn the mixture screw's counter-clockwise (richer) until a loss of engine R.P.M. is indicated on tachometer. Then turn the mixture screws clockwise (lean) until engine R.P.M. again starts to decrease. If the idle speed changes more than 30 R.P.M. during adjustment, repeat the adjustment.

STEP ELEVEN
INSTALLING THERMAL SWITCH

● Place electric thermostat switch on underside of upper radiator hose near radiator, being careful not to contact hose clamp or radiator. Place cellulose rubber pad over switch, and wrap switch and pad tightly to radiator hose with tape provided. **SPECIAL NOTE: Thermal Switch should be completely covered so that it is not exposed to any air flow. THERMAL SWITCH SHOULD BE INSTALLED LAST IN ALL APPLICATIONS.**

STEP TWELVE
CHECKING THE SYSTEM

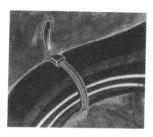

● Using the tie straps supplied, secure wire to radiator hose so as not to interfere with fan or belts. Secure other wires or straps furnished. Check all connections.

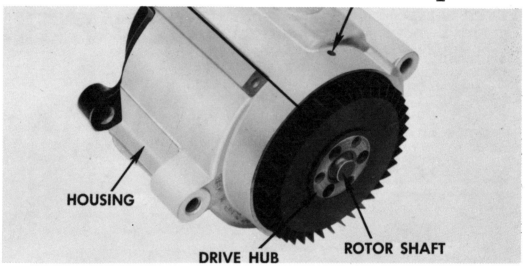

HOUSING

DRIVE HUB   ROTOR SHAFT

AIR PUMP

The first air pumps, used as original equipment on 1966 and 1967 California cars, had 3 vanes or "paddles" that rotated and moved the air from the inlet to the outlet. These pumps were repairable and Saginaw Division of General Motors supplied the parts. In 1968, the 2-vane pump came out and it was considered a replacement item, with internal parts not available. A few months later, the production of replacement parts for the 3-vane pump was discontinued. Since then, the only way to fix a worn-out pump, either 2-vane or 3-vane, is to replace it.

Some repairs are allowed on the outside of the pump, such as replacing the filter fan, the relief valve, or an exhaust tube, but there are more things that you can't do to the pump which you should know about. The following is a list of definite "don'ts."

Air Pump Service Precautions:

1. Do not clamp the pump in a vise.
2. Do not pry on the aluminum housing to adjust the drive belt tension.
3. Do not use hammer to install a relief valve pressure setting plug.
4. Do not exceed 25 ft lbs. torque on the pump mounting bolts.
5. Do not disassemble the pump or remove the rear cover.
6. Do not remove the drive hub when replacing the filter fan.
7. Do not replace the filter fan by driving or hammering.
8. Do not clean the filter fan. Replace it, if it is clogged.
9. Do not replace the pump if it squeaks when turned by hand.
10. Do not lubricate the pump.
11. Do not run the engine on the road with the pump belt removed.

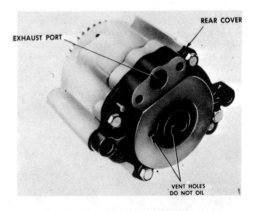

REAR COVER

EXHAUST PORT

VENT HOLES
DO NOT OIL

The design of the mounting ears and the exhaust outlet varies with different applications, but the basic pump is still the same. The rear cover is cast iron. If you must pry on the pump do it on the rear cover

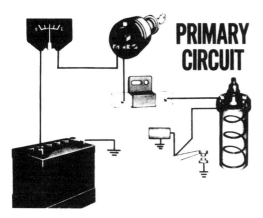

PRIMARY CIRCUIT

The primary circuit is the path taken by battery current. The ammeter shows in this Chrysler system, because their ignition switch connects there. The resistor by-pass is not shown

12. Do not try to keep the pulley from turning by inserting tools into the filter fan.

After reading the list, you may feel that there isn't much you can do to the pump. The feeling should be encouraged, even on the old 3-vane pump. One of the reasons Saginaw discontinued replacement parts was due to the difficulty that most mechanics had getting the pump back together. If you are still tempted to take a 2-vane pump apart, read this statement direct from Saginaw. "The air injection pump is a positive displacement vane type which is permanently lubricated and requires no periodic maintenance." The only thing the pump does require is a periodic test to see if it is working, an inspection of the filter fan for dirt or damage, and a check of the relief valve, if used.

How Ignition System Works

The ignition system starts at the battery with only 12 volts and ends at the spark plug, where there are 20,-000 volts or more. This seemingly miraculous transformation of electricity from simple battery voltage into the kind of voltage you find in overhead power lines is accomplished by the ignition coil. Current from the car battery is put into the coil and it trans-

forms to the higher voltage needed to jump the gap at the spark plug. Its principle is not much different from a toy train transformer, which works the opposite way, cutting down the 110 volt house current to a safe 6 or 12 volts for train operation.

PRIMARY CIRCUIT

The ignition system is divided into two circuits: the primary, which conducts the car battery current; and the secondary, which conducts the high voltage. The primary circuit starts at the battery and goes to the ignition switch, through the coil primary windings and then to ground at the distributor points.

SECONDARY CIRCUIT

The secondary circuit is a series of windings inside the coil, separate from the primary windings. The secondary windings are connected to the coil tower. From there, the circuit is through the coil wire to the center of the distributor cap, through the distributor rotor to the cap electrodes and then down the spark plug wires to the plugs.

COIL ACTION

When the distributor points are closed, completing the circuit to ground, current flows through the primary circuit. As the current flows

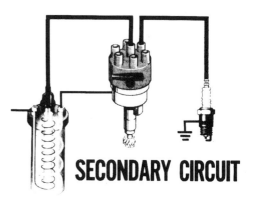

SECONDARY CIRCUIT

The secondary circuit starts at the coil and ends at the plug. It includes the cap and rotor

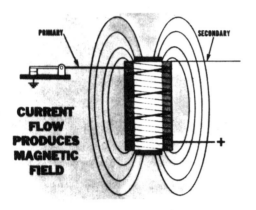

When the points are closed, current flowing through the primary coil windings produces a magnetic field

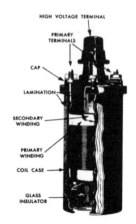

Refinements inside the coil include a particular way of wrapping the windings, and laminations to increase the field strength

through the coil primary windings, it creates a magnetic field around and in the coil. This field surrounds the secondary windings in the coil.

When the distributor points open, the magnetic field collapses, and a high voltage is created in the coil secondary winding. This high voltage tries to get out of the coil anyway it can, but electricity is inherently "lazy," so that it takes the easiest path, which is along the coil lead to the distributor and out to the spark plugs.

To review then: When current running through the coil primary circuit is interrupted, it creates a spark in the coil secondary circuit. That's really all there is to a basic ignition system. Exactly why the electricity does this, is something we should leave to the scientists. If we just accept the ignition system's basic principles, we can work with it, repair it,

and keep it in good shape, without going into a lot of unnecessary electrical theory. Of course, there are refinements to the basic systems, so let's cover some of those next.

CONDENSER

If you have ever tried to start a car without a condenser, you know there is more to the primary circuit than just a coil and a pair of points. When the points first open, the current flowing through the coil primary has a tendency to keep going. This continuing flow of current happens just as the points start to separate. It doesn't continue for very long, but long enough that instead of getting a clean break of current, it dwindles down and does not create a good spark. This is where the condenser comes in.

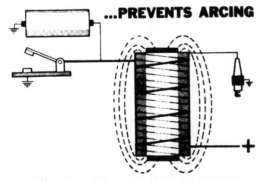

When the points open, the field collapses, creating a high voltage in the coil secondary, which jumps the gap at the plug

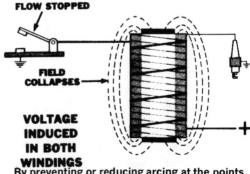

By preventing or reducing arcing at the points, the condenser increases the amount of voltage that the coil can put out. Without the condenser, the car won't run

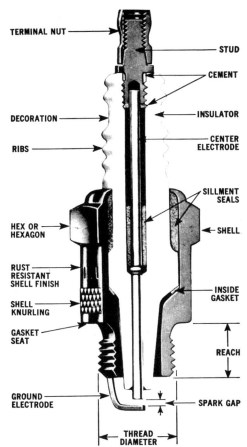

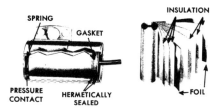

A condenser must be precision made if it is going to last. This quality Delco condenser is typical

A basic Champion spark plug. There are many variations from this basic design, including booster gaps, extended noses, resistor types, etc.

The condenser lead is connected to the insulated point. The can of the condenser is grounded. The condenser lead is not electrically connected to the condenser can. The condenser is made of two rolls of wound tin foil which are very close to each other, but with insulating paper between them. One of the rolls is connected to the can and the other is connected to the lead, or pigtail. When the points start to open, the condenser acts as a storage place for the current that would try to keep flowing across the points. Current is "lazy" and it will always follow the easiest path. In this case, the attraction of the condenser is much greater than the attraction of the separating points. So, the condenser attracts the current that is still flowing and we get a clean break at the points, which gives a maximum buildup of voltage in the coil.

If you don't believe that a condenser actually attracts electricity and stores it, try charging one sometimes and then touching it to your arm. You'll get quite a shock, but nothing that will hurt you. While the points are separated, the condenser remains charged, but when the points close again, this discharges the condenser so that it is ready the next time the points open.

A condenser has a secondary function and that is to absorb the voltage that is induced in the primary circuit when the points separate. This voltage that is induced in the primary circuit is unwanted, but it happens and there is nothing that can be done about it, except to try to get rid of it. The induced voltage amounts to about 250 volts. If it is not eliminated, it will tend to arc across the contact points and wear them out very fast. The condenser absorbs this voltage and thereby the points last much longer.

It's easy to see why a shorted condenser would keep a car from running. The current flowing through the primary circuit would go right through the condenser to ground and the opening of the points would not stop it.

SPARK PLUGS

Spark plugs do not make sparks, although many people think that they do. A spark plug is nothing more than a stationary pair of points inside the combustion chamber. The center electrode is insulated and the side electrode is connected to ground.

High voltage from the coil flows down the spark plug wire into the center electrode. If the coil has enough capacity, the voltage will jump across the gap, creating a spark.

A strange thing about the automotive ignition coil is that it only creates enough voltage to jump the gap. If you hook a spark plug up so that the spark jumps the gap while the plug is lying on the fender, it only takes a few thousand volts. But, if you put the plug inside the engine, and subject it to the compression and the heat, then it's a different story. Under full-throttle, it can require 20,000 volts or more to jump across that little gap. If the coil puts out 25,000 volts, but the engine only requires a maximum of 20,000 volts, then you are in good shape and you'll never have a misfire because of insufficient voltage.

The center electrode of the spark plug, being insulated from ground by the porcelain that surrounds it, is sensitive to deposits on that porcelain. If enough oil gets on the porcelain, then the current will flow from the center electrode right down the side of the porcelain to ground without jumping the gap. Of course this results in a dead spark plug and a dead cylinder and maybe a dead engine, if enough of the plugs are in this condition. That is what a fouled plug is; one which has lost its insulation between the center electrode and ground because of deposits. Clean the deposits off, or put in a new plug that doesn't have deposits and the voltage will once again be able to jump the gap.

Spark plugs do not last forever. Because of the deposits they either have to be cleaned or replaced at regular intervals. Some cars driven with just the right amount of full-throttle running to keep the plugs clean, can get fantastic life out of a set of plugs. In fact, the plugs may last longer than the points. That is one of the weaknesses of the modern ignition system, that the points do not last. The main thing that destroys the points is the heavy current that passes across them just as they start to separate. A small arc occurs every time the points open resulting in metal being deposited from one point face to the other, building up a hill and valley. The condenser helps, but there is no way with the basic ignition system to eliminate this entirely, and so sooner or later the points are

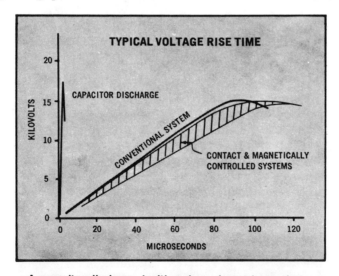

A capacitor discharge ignition rises almost instantly to a voltage high enough to jump the plug gap. The spark will occur before the voltage gets a chance to leak away on a partially fouled plug. Because of this, a good CD ignition will fire plugs that won't even run an engine with a conventional ignition

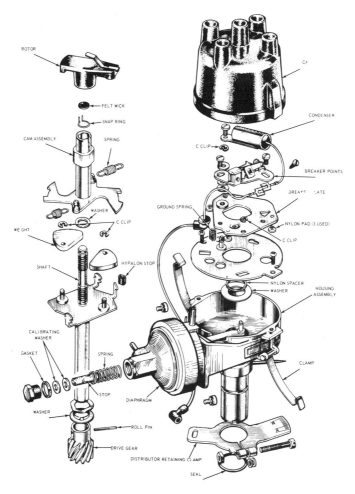

Exploded view of 4 cylinder distributor—1972 Ford

going to have to be replaced or the engine won't run.

TRANSISTOR IGNITION

Transistor ignition is supposedly designed to eliminate problems with the points. The heavy switching current is taken care of by one or more transistors and all the points do is carry a very light current that turns the transistors on and off. Unfortunately, some transistor ignition systems have more sales engineering behind them than technical know-how. If the systems are designed so that the points carry a very small amount of current, other problems arise. It's impossible to keep the oil vapor and fumes out of the distributor and sooner or later the points are going to be coated with engine fumes. The re-

sult, with the poorly designed transistor ignition systes, is that instead of replacing the points because they are pitted, now you have to clean them because they are corroded. The better transistor ignitions pass enough current through the points to burn off deposits, but not wear them out prematurely.

CAPACITOR DISCHARGE IGNITION

Another development in the search for longer life and better spark is the capacitor discharge ignition. Instead of relying on the collapse of the primary field in the ignition coil, the capacitor discharge ignition charges up a large capacitor or condenser. When the points separate, they trigger electronic circuitry that fires the high

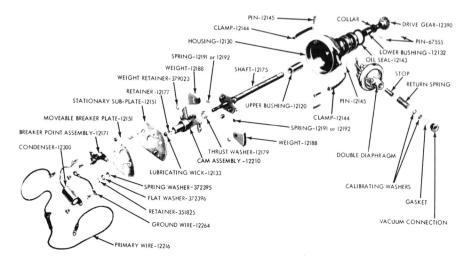

PIN-12145
COLLAR
DRIVE GEAR-12390
CLAMP-12144
PIN-67555
HOUSING-12130
LOWER BUSHING - 12132
SPRING-12191 or 12192
OIL SEAL-12143
WEIGHT-12188
SHAFT-12175
STOP
WEIGHT RETAINER-379023
RETURN SPRING
RETAINER -12177
STATIONARY SUB-PLATE-12151
UPPER BUSHING-12120
PIN -12145
MOVEABLE BREAKER PLATE-12151
CLAMP-12144
BREAKER POINT ASSEMBLY-12171
SPRING-12191 or 12192
CONDENSER-12300
WEIGHT-12188
THRUST WASHER-12179
DOUBLE DIAPHRAGM
CAM ASSEMBLY - 12210
LUBRICATING WICK-12133
SPRING WASHER-372395
CALIBRATING WASHERS
FLAT WASHER-372396
GASKET
RETAINER-351825
GROUND WIRE-12264
VACUUM CONNECTION
PRIMARY WIRE-12216

V8 distributor assembly—1971 Ford

voltage in the capacitor through the coil primary. This creates a much higher voltage than could be created with the simple collapse of the primary field. Capacitor discharge ignition seems like the best way to go when you first think about it, but the problem is that if it is made too hot, the spark tries to go to ground everywhere except across the spark plug. What capacitor discharge systems really need is a much larger distributor, designed especially so that everything is twice as far away as it is in the present distributors.

PULSE CONTROLLED DISTRIBUTOR

The ultimate in either capacitor discharge or transitor ignitions is to do away with the points entirely. That way you eliminate the weakest part of the ignition system and all you have to worry about from then on is when to change the spark plugs. Aside from the problem of changing the points, another advantage of an ignition system without points is that there is no point rubbing block to wear and gradually retard the spark as mileage builds up. If an ignition without points is set to a certain initial timing, it will remain that way for the entire life of the car. The only thing that could possibly retard the spark is wear on the camshaft timing chain or on the distributor drive gears, which is very small.

Pointless ignitions use a magnet mounted inside the distributor to sense the position of the distributor shaft. Mounted on the shaft are a series of little bumps or projections, one projection for each cylinder in the engine. As the projections rotate on the distributor, the magnet senses their position because the magnetic field increases when each projection gets opposite the magnet. Some systems use a ring magnet that picks up magnetic field from all the projections at once.

SPARK TIMING

If an engine ran at constant speed and load, you would never have to worry about varying the spark timing. It could be locked in position for the best setting at that speed and load. But an engine operates from a few hundred rpm at idle to 4,000 rpm and over at top speed. Load varies, too, according to throttle opening. At any given speed and load, there is only one timing setting that is best for maximum horsepower and to prevent engine damage.

Timing must be precise so that the burning mixture in the combustion chamber exerts its maximum push on the piston. The mixture does not explode, it burns at a definite rate that can be measured in the laboratory. A certain amount of time passes after the spark ignites the mixture, until it reaches maximum burning rate and gives its biggest push to the piston. Another factor is the movement of the piston. The burning mixture must push on the piston when it is moving down in the cylinder. If the mixture burns too early, it will push on the piston before it reaches top dead center (TDC), which will make the engine run backwards. If the mixture burns too late, the piston will be already near the bottom of the stroke and the power of the burning mixture will be lost.

CENTRIFUGAL ADVANCE

The burning time of the mixture

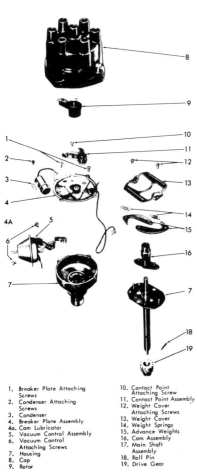

1. Breaker Plate Attaching Screws
2. Condenser Attaching Screws
3. Condenser
4. Breaker Plate Assembly
4a. Cam Lubricator
5. Vacuum Control Assembly
6. Vacuum Control Attaching Screws
7. Housing
8. Cap
9. Rotor
10. Contact Point Attaching Screw
11. Contact Point Assembly
12. Weight Cover Attaching Screws
13. Weight Cover
14. Weight Springs
15. Advance Weights
16. Cam Assembly
17. Main Shaft Assembly
18. Roll Pin
19. Drive Gear

L-6 distributor—exploded view

changes very little, but the speed of the piston changes a lot. As the piston goes faster, it is necessary to make the spark occur sooner, so that the push on the piston will still happen as it starts its downstroke. This advancing of the spark according to engine speed is done by the centrifugal advance mechanism. The distributor cam is moved by certrifugal weights, opposed by small springs. As the engine goes faster, the weights fly out and move the cam in the direction of distributor rotation to advance the spark. As the engine slows down, the springs pull the weights back, and the cam is moved to the neutral position. Some distributors have the weights built into the bowl of the distributor, under the breaker plate. Others have the weigths at the end of the shaft, under the rotor, where they are easily inspected and lubricated.

Centrifugal advance is set on a distributor machine, by changing the small weight springs, or by changing the weights themselves. Centrifugal advance curves or checking points are listed by the car makers or distributor machine manufacturers, so that distributors can be set according to factory specs.

VACUUM ADVANCE

When the throttle is wide-open, a full charge enters the combustion chamber, and the engine will put out as much horsepower as it can, according to where the spark timing is set. The most horsepower is obtained by advancing the spark until cylinder pressures are so high that the fuel is close to the detonation point. But that is dangerous, because detonation cracks pistons and destroys engines. So, the spark is kept retarded to a safe setting.

At part-throttle, vacuum exists in the combustion chamber and the mixture is thin. This thin mixture does not produce as much heat when it burns, so that it does not detonate as easily. The spark can be advanced to increase cylinder pressure and get more push out of each cylinder, without fear of destroying the engine.

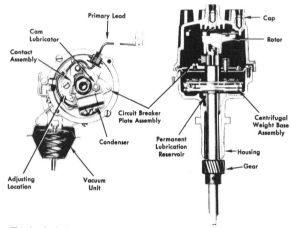

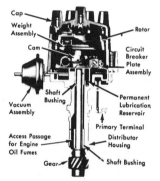

Typical late model American Motors distributor —6 cylinder

Typical late model American Motors distributor—8 cylinder

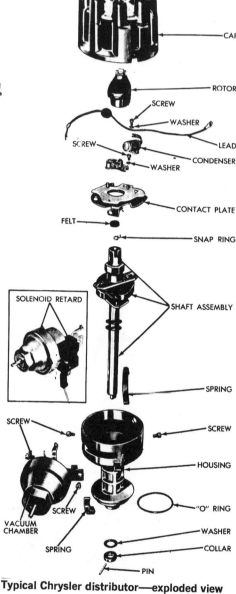

Typical Chrysler distributor—exploded view

The vacuum advance unit gives exactly what is needed in spark advance according to engine load. When the throttle is wide-open, there is no vacuum, so the vacuum advance does nothing and the spark stays retarded. When there is vacuum, as at part-throttle cruising, the vacuum advance unit advances the spark to get more part-throttle horsepower and thereby better gas mileage. The vacuum advance changes the spark timing by moving the breaker plate and the distributor points against the direction of distributor cam rotation to advance the spark. When there is no vacuum, a spring in the advance unit moves the breaker plate back to the no advance position. All vacuum units can be checked by running the distributor on a distributor machine which has a vacuum pump. Adjustable units can be calibrated on the machine, but non-adjustable ones must be replaced if they don't fall within the range specified by the manufacturer.

INITIAL ADVANCE

Vacuum advance and centrifugal advance take care of spark timing while the engine is operating above idle, but there has to be a starting point set when the distributor is installed in the engine. This starting

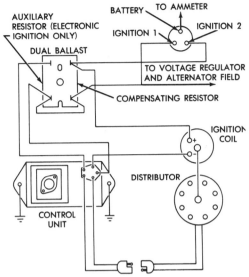

Chrysler Electronic Ignition 8 cylinder wiring diagram

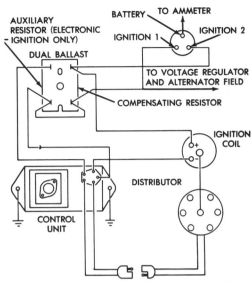

Chrysler Electronic Ignition 6 cylinder wiring diagram

point, called the "initial advance" or "initial timing," is set by moving the distributor in its mounting while the engine is idling, until the spark occurs according to the manufacturer's specification. Initial advance is important because if it is off 1°, then the centrifugal advance will be off 1° throughout the entire speed range of the engine. Initial advance is always set with the vacuum advance line disconnected, to be sure that the vacuum advance unit has returned to the neutral position. Engine rpm, when setting initial advance, must be low enough that the centrifugal advance weights have not started to come out and advance the timing. Some car makers actually specify an idle speed for setting the initial timing lower than the curb idle speed. On those cars, you reduce the idle speed to set the timing and then you have to set it back up again after you are through.

TOTAL ADVANCE

The total amount of spark advance that you can read off the front pulley, at any given engine speed, is the total advance reading. Total advance is merely a checking setting, usually given as a range of about 30 = 40° at 2,500 engine rpm. No setting is ever made on the engine in order to get a

Close-up of the distributor used with the Chrysler Electronic Ignition. Note the single screw and screwdriver slot for adjusting the air gap

specified amount of total advance. Vacuum advance and centrifugal advance are calibrated into the distributor on a distributor machine. The initial advance is then set when the distributor is installed in the engine. Total advance is merely a quick check, used to find out if the total amount of advance read off the front pulley is falling in an acceptable range. Total advance is usually checked with a timing light that has a degree meter. It is possible to check total advance without one of these special timing lights, but then you have to estimate the distance that the mark moves on the front pulley and this can be difficult because the degree marks usually do not go that far.

TRANSISTOR IGNITION

Transistor ignitions of various designs have been used since approximately 1962. The following descriptions are of the original equipment systems that have been in use since approximately 1966, listed alphabetically.

Chrysler Electronic Ignition

The Chrysler system first appeared in 1971 as a mid-year change on the 340 cu in. engine with manual transmission. In 1973, the Chrysler electronic ignition system was used on all Chrysler Corporation vehicles built in the United States. The Chrysler system uses a standard coil, distributor housing, rotor and cap. With the cap on the distributor, it looks the same as the standard model, except that there are two wires going into the distributor body instead of one. The easiest way to recognize the Chrysler electronic ignition is by the dual ballast resistor. One of the resistors is the normal coil resistor and the other is a special resistor to prevent excess current from damaging the electronic control unit.

Inside the standard distributor housing, Chrysler has replaced the ordinary cam with what they call a "reluctor." This reluctor has points on it which are much more prominent than the bumps on an ordinary cam. Taking the place of the distributor points is a pickup unit, which is a permanent magnet with a coil of wire which senses the increased magnetism when one of the reluctor points comes near the magnetic pole piece. The gap between the pole piece and the reluctor points is adjustable, just about the same way that the distributor points are adjustable. If a new pickup unit is installed, the gap should be set to 0.008 in., with a feeler gauge. However, since this is an air gap with no rubbing contact, it does not change in service and there is no need to check it unless you feel that somebody has been tampering with it. The pickup unit can be bent if you are not careful. Therefore, if you do set the gap, be careful not to bend anything out of shape. Because of the magnetism on the pickup, you will find that the ordinary steel feeler gauge will drag so much that you can't feel the gap. It is necessary to use a non-magnetic feeler gauge of some kind or a piece of brass shim stock 0.008 in. thick. The reluctor points are deliberately made with sharp edges that look as if they are unfinished or rough. This is done to get a quick cutoff of the magnetic field when the reluctor point passes the pole piece. Do not dress or file the reluctor in any way. If you do, you will probably ruin it and cause erratic sensing. The gap between the

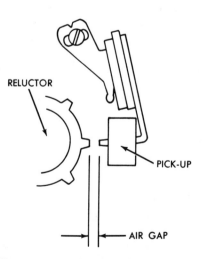

Air gap is set with a non-magnetic feeler gauge between the pick up coil and the reluctor

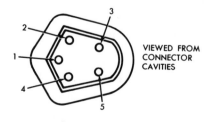

Follow this numbering of the connector cavities when performing electrical tests

pole piece and the reluctor does not effect dwell. In fact, a dwell meter cannot be used on the electronic system.

The ballast resistor has two units mounted in a ceramic block on the firewall. One of the resistors has a value of 0.55 ohms and it is the regular resistor that goes between the ignition switch and the coil primary circuit. You can identify this regular resistor because it is exposed on the back side of the ceramic block. The other resistor, built into the same ceramic block, has a resistance value of 5 ohms. This resistor can be identified easily because it is completely covered up with insulating cement on the back side of the ceramic block.

TESTING THE CHRYSLER SYSTEM

The easiest way to test the Chrysler electronic ignition is with the Chrysler electronic ignition tester. It is available from Miller's Special Tools, 32615 Park Lane, Garden City, Michigan 41835. Or you may be able to purchase one through your local Chrysler Corporation dealer. The tester is very easy to use and comes with instructions on the back. You simply plug it into the system and lights come on to tell you what is good and what is bad. The tester has been made in two versions: No. C-4166 is the early unit that was used to check 1971 and 1972 systems. To check 1973 systems, you must have an adapter No. C-4166-1 that plugs into the tester. The adaptor can be left in place because it can be used to test 1971 and 1972 units also. The latest tester is No. C-4166-A. It comes without an adapter because it has the capacity built into the box to test units of any year.

The Chrysler system can be tested without the special tester, although not as easily. In case of trouble, the first and easiest thing to check is the air gap which should be 0.008 in. When checking the air gap, you must be careful not to force the feeler gauge between the pickup and a reluctor point. Use as much caution in setting gap on an electronic ignition as you would in setting gap on an or-

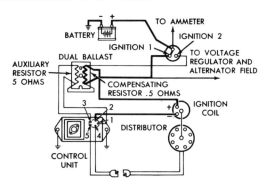

Chrysler Electronic Ignition test No. 1

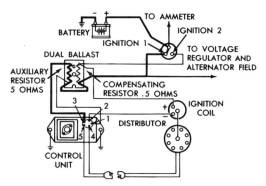

Chrysler Electronic Ignition Test No. 2

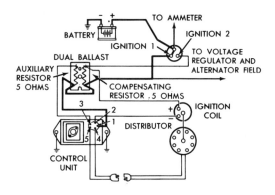

Chrysler Electronic Ignition Test No. 3

dinary set of distributor points. After checking the gap, check the wiring for a bad connection, loose connections, or breaks in the wire. The next step is to be sure that the ignition switch is off. Then remove the connector at the control unit. This connector has a phillips head screw in the middle that must be unscrewed all the way before the connector will come off. The connector must never be removed while the ignition switch is on. If you do, the control unit will

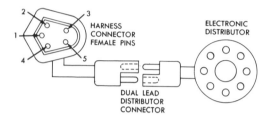

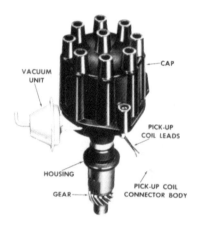

Chrysler Electronic Ignition Test No. 4

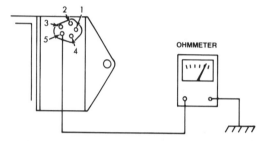

Chrysler Electronic Ignition Test No. 5. Note that this test is made on the unit itself

The cap is usually colored red on a magnetic pulse system, but you can also spot the system by seeing the connector and two wires coming out of the distributor

probably be damaged.

Refer to the diagram for the proper numbering of the holes in the connector cavities. Note that we are making test connections to the holes or cavities in the *connector* that you just removed from the control unit.

Test No. 1 Turn the ignition switch on and connect a voltmeter with the negative (−) lead to a good ground and the positive (+) lead to connector cavity No. 1. The voltage reading should be within one volt of battery voltage, with all accessories off. If there is more than one volt difference, refer to the diagram labeled "Test One." The heavy black line shows the circuit to connector cavity No. 1 and you must check that circuit to find the reason for the loss in voltage. It is probably a bad connection or a broken wire.

Test No. 2 With the ignition switch on, connect the positive (+) voltmeter lead to connector cavity No. 2 and the negative (−) lead to a good ground. With all accessories off, the voltage reading should be within one volt of battery voltage. If there is more than one volt difference, refer to illustration "Test Two." The heavy black line shows the circuit that must be checked to find the lost voltage.

Test No. 3 With the ignition switch on, connect the positive voltmeter lead to the connector cavity. No. 3 and the negative (−) voltmeter lead to a good ground. Your voltage reading should be within one volt of battery voltage with all accessories off. If there is more than one volt difference, refer to the illustration labeled "Test Three." The heavy black line shows the circuit that must be checked to find the lost voltage.

Test No. 4 Turn the ignition switch off and connect an ohmmeter between the connector cavities No. 4 and No. 5. The resistance reading should be between 350 and 550 ohms. If the reading is outside that range, you should disconnect the dual plug in the two wires coming out of the distributor. Use the ohmmeter on the plug to measure the resistance of the pickup coil in the distributor. If the reading does not fall between 350 and 550 ohms, the pick-up coil is defective and must be replaced. If the reading is within the range now, but was not in the range when you measured it at the harness connector, it means that the wiring harness from the distributor connector to the control unit plug has a break in it or a defective connection. To check it, use the ohmmeter between either one of the harness pins on the dual connector and ground with the connector off the control unit. It should show an open circuit. If the ohmmeter indicates continuity,

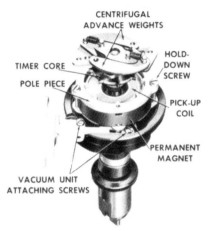

The pole piece and timer core completely encircle the timer core. The number of points on the pole piece or timer core must be the same as the number of cylinders

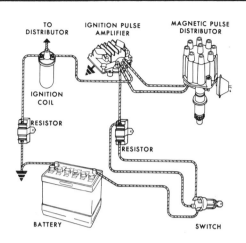

A typical wiring diagram for the magnetic pulse ignition. Note that the primary resistor is in the ground circuit from the coil

it means that the wiring is grounded. The same test can be made on the distributor pickup unit. If continuity shows between either one of the two leads and ground, it means that the pickup unit or the wires are grounded and will have to be replaced.

Test No. 5 This is the only test that is performed on the control unit itself, in the absence of the special tester. With the control plug disconnected, hook up an ohmmeter to ground and to control unit pin No. 5. The ohmmeter should show continuity. If you get a reading of infinity, indicating no continuity, try to get a better ground on the control unit by tightening the bolts that hold it to the firewall. If continuity still does not show, the control unit is defective and must be replaced.

CAUTION: *Do not use battery powered test lights or non-powered test lights with the vehicle battery to make this test on the control unit. Battery voltage, if accidentally applied to the wrong pin on the control unit, could ruin it.*

Test No. 6 Put all plugs back in their normal running position, turn the ignition switch on and remove the coil lead from the center of the distributor. Hold the lead approximately 3/16 in. from ground on the engine and crank the engine. If you do not

get a spark off the end of the wire, replace the control unit. Make the test again, after replacing the unit. If you still do not get a spark, replace the ignition coil. When replacing the control unit temporarily, remember that it must be grounded.

Because the ignition coil with this system is a standard coil, it can be tested on any available coil tester. However, make sure that you disconnect the wires from the coil, so that the coil tester cannot do any damage to the electronic control unit.

Delco Remy Magnetic Pulse Ignition

Delco Remy Magnetic Pulse Ignition was used for almost ten years as original equipment on certain models, starting back in about 1962. The system uses a special distributor, a separate amplifier, a special coil and two resistors mounted in ceramic blocks. The distributor has two primary wires leading to a pickup coil which surrounds the distributor shaft. A permanent magnet and a pole piece with one projection facing inward for each cylinder in the engine are also mounted so that they encircle the distributor shaft. Taking the place of the distributor cam is a timer core with one projection on it for each cylinder in the engine. When the projections on the timer core line up with the projections on the pole piece, in-

The amplifier looks very much like the transistor generator regulator used in the same years. The regulator has an adjustment screw in plain sight, but the amplifier doesn't

creased magnetism flows through the unit. The pickup coil senses this increased magnetism, which tells the amplifier when to collapse the field in the coil to create the spark. One of the resistors used is the normal ballast resistor and the other is connected between the coil primary negative terminal and ground. On some applications, these resistors may be a wire type that is hidden inside the wiring harness instead of being mounted in a ceramic block.

Troubleshooting should start at the pickup coil, by disconnecting it and taking a measurement of the resistance of the coil with an ohmmeter. The resistance should be 550-650 ohms. If the reading is infinity, the coil is open and if the reading is below the specifications, the coil is probably shorted. A check for a ground in the pickup coil should be

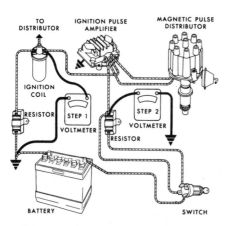

Voltmeter checks of the magnetic pulse system

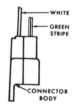

Use this diagram to check for the correct position of the leads in the connector near the distributor

made by connecting the ohmmeter lead to each pickup coil wire, in turn, and also to the distributor housing. If the pickup coil checks okay, then the ignition coil should be tested on a coil tester. You must be sure that the tester is capable of testing this special coil. If nothing is wrong with the coil, then a new amplifier should be substituted to see if it corrects the problem.

ENGINE SURGING

Severe engine surging may be caused by the two distributor leads being reversed in the connector or by an intermittent open circuit in the distributor pickup coil. Refer to the illustrations to see where the white and green stripe leads should be in the connector body. Because the pickup coil is moved by the vacuum advance unit to advance the spark, it's possible that the constant flexing of the wires to the coil could break them. If the surging stops when the vacuum unit is disconnected, this is a good indication that the pickup coil wires are defective. You can also check for this condition by using an ohmmeter on the pickup coil and moving the vacuum advance unit with a hand vacuum pump. If the reading changes as the coil moves, it means the wires are faulty.

VOLTAGE CHECKS OF THE CIRCUIT

With the ignition switch on and the engine not running, connect a voltmeter to the coil primary circuit positive (+) terminal and to ground. The voltage reading should be 6-7 volts. If you get full battery voltage, it indicates an open circuit in the coil primary or in the wiring or resistor be-

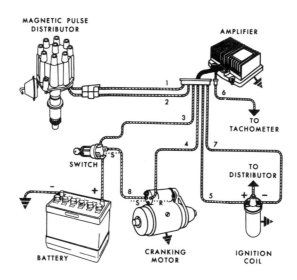

Typical wiring circuit with 6-terminal connector on amplifier wiring harness

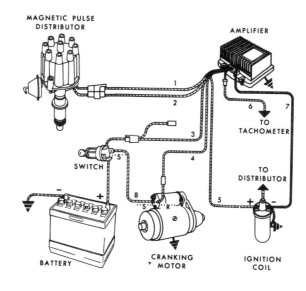

Typical wiring circuit when amplifier does not use a 6-terminal connector in the harness. The No. 7 lead represents a shield over the No. 5 lead

tween the coil primary and ground. If the reading is zero, there is an open or bad connection between the battery and the coil, which includes the ballast resistor and ignition switch.

VOLTAGE CHECKS STEP TWO

Voltage can also be checked at the amplifier side of the ballast resistor with the ignition switch on. If the ballast resistor is the resistance wire type which is inside the wire in the

harness, disconnect the plug at the amplifier and make the connection as shown in the illustration. Voltage at the amplifier side of the ballast resistor should also be 6-7 volts. If you get a zero reading, there is an open circuit between the battery and the ballast resistor. If you get full battery voltage, there is an open circuit between the resistor and the ignition coil. If the voltage checks are satisfactory, it indicates that the trouble

Voltmeter check of the harness connector

must be in the amplifier. A new amplifier should be substituted to see if it cures the problem.

Delco Remy Magnetic Pulse Capacitor Discharge Ignition

Delco Remy's capacitor discharge ignition (CDI) uses a magnetic pulse distributor, a separate pulse amplifier, and a special coil. The CDI amplifier has totally different circuitry from the amplifier used with the normal magnetic pulse ignition. The CDI amplifier charges a capacitor to about 300 volts and then fires it through the coil primary to make the spark.

The capacitor discharge amplifier may be identified with either the letters "CD" (capacitor discharge) or "UHV" (ultra high voltage).

TROUBLESHOOTING THE DELCO REMY
CDI SYSTEM

Poor engine performance will show up as one of the following conditions:
1. The engine will not start at all.
2. The engine will start, but will not run.
3. The engine will miss or surge.

If any of the above conditions are present, then the applicable troubleshooting procedure should be performed, as outlined below.

CAUTION: *When troubleshooting the Delco Remy CDI system, be careful to avoid accidentally shorting or grounding any of the system components; doing this, even for an instant, could damage the amplifier and wiring.*

ENGINE WILL NOT RUN

1. To determine if the problem lies in the ignition system, remove and hold one spark plug wire about ¼ in. away from the block. Crank the engine.
2. If a spark occurs, the trouble

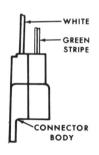

The connector near the distributor must have the leads in this position

probably is not in the ignition system.
3. If there is no spark and/or if the fuel system checks out okay, the ignition system should be checked.
4. The spark plugs, wiring, distributor cap and coil tower can be checked in the usual way.

NOTE: *The special coil used with this CDI system cannot be tested with a standard coil tester. A normal voltmeter will read zero volts, when connected across the ignition coil with the ignition turned on.*

5. The ignition coil can be checked for both primary and secondary winding continuity, by using an ohmmeter in the following manner:
   a. Disconnect the leads from the coil.
   b. Connect the ohmmeter across the primary terminals. If a reading of infinity is obtained, the primary circuit winding is open.
   c. Connect the ohmmeter across the center tower and the coil's case. If a reading of infinity is obtained, the secondary circuit winding is open.

NOTE: *Use the middle or high resistance range on the ohmmeter to check the secondary windings.*

6. Replace the coil, if it fails either of the above tests.
7. If the coil is okay, go on to the next test, after reconnecting its leads.

Amplifier

1. Temporarily connect a jumper

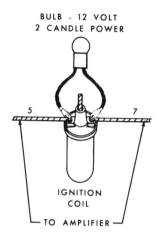

BULB - 12 VOLT
2 CANDLE POWER

5   7

IGNITION
COIL

TO AMPLIFIER

Testing with a bulb and leads for the cause of trouble

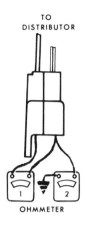

TO
DISTRIBUTOR

1   2

OHMMETER

Distributor checks with an ohmmeter

lead between the amplifier housing to a good ground. If the engine will not start and run, the amplifier is not properly grounded. Correct as required.

2. Connect a 12-volt, 2-candlepower bulb to the primary terminals of the ignition coil.

3. Crank the engine.
   a. If the bulb flickers on and off, the amplifier is operating properly. In this case, recheck the secondary system for the cause of the "no run" condition.
   b. If the bulb does not flicker on and off, proceed to next test.

Distributor

1. Insure that the two distributor leads are connected to the distributor connector body as illustrated.

2. With the distributor connector disconnected from the harness connector, connect an ohmmeter (1) to the two terminals of the distributor connector as shown.

3. Connect a vacuum source to the distributor, and observe the ohmmeter reading throughout the vacuum range. (The distributor need not be removed from the engine.)

4. Any reading outside the 550-750 ohm range indicates a defective pickup coil in the distributor.

5. Remove one ohmmeter (2) lead

from the connector body and connect to ground.

6. Observe the ohmmeter reading throughout the vacuum range.

7. Any reading less than infinite indicates a defective pickup coil.

8. Reconnect the harness connector to the distributor connector.

Continuity Checks

There are two different procedures, depending upon the type of wiring harness used. The first test applies to the first type of wiring harness illustrated, and the second test to the second illustration.

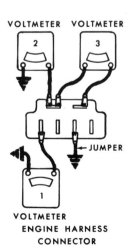

VOLTMETER   VOLTMETER

2   3

JUMPER

VOLTMETER
ENGINE HARNESS
CONNECTOR

Voltmeter continuity tests on a system with a 6 terminal connector

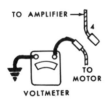

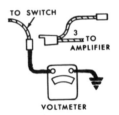

Voltmeter test for system without 6 terminal connector

To Test Type 1:

1. Separate the harness and amplifier connections.
2. Attach a voltmeter to the harness connector as illustrated in "test 1", and turn the ignition switch to "START."
3. If there is no reading, check for an open circuit in the No. 4 lead.
4. Next, connect the voltmeter as shown in "test 2", and turn the ignition switch to "IGN" or "RUN."
5. If there is no reading, check for an open circuit in the No. 3 lead.
6. Then attach a voltmeter and a jumper lead as shown in "test 3", and leave the ignition switch in the same position as in Step 4.
7. If there is no reading, check for an open circuit in lead No. 5 or lead No. 7, or for one in the ignition coil primary circuit (see above).
8. If all of the leads check out okay, then replace the amplifier.

To Test Type 2:

Carefully inspect wiring connections to ensure that they are clean and tight. If satisfactory, disconnect the amplifier No. 3 and No. 4 leads from the two connectors. Proceed as follows:

1. Connect a voltmeter from ground to the No. 4 connector lead.
2. Turn the switch to "Start" position.
3. If the reading is zero, the circuit is open between the connector body and the battery.
4. If a reading is obtained, connect the voltmeter from ground to the No. 3 connector lead.

Voltmeter test for system without 6 terminal connector

5. Turn the switch to the "run" position.
6. If the reading is zero, the circuit is open between the connector body and the switch.
7. If a reading is obtained, replace the amplifier.

ENGINE WILL START BUT WILL NOT RUN

If the engine starts but then stops, as soon as the ignition switch is returned to the "RUN" position, perform the following tests:

1. For the first type of wiring harness illustrated above, separate the amplifier connector from the harness connector and connect a voltmeter as shown in "test 2".
2. For the second type of harness, illustrated above, perform the following:
   a. Be sure that the leads are securely connected to the No. 3 lead body.
   b. If the connection is satisfactory, connect a voltmeter from ground to the connection side of the connector.
3. For both types of harness, turn the ignition switch to "RUN."
4. If there is no reading, an open circuit exists between the voltmeter connection and the ignition switch. Trace and fix.
5. If there is a reading, the fault must lie in the amplifier; replace it.

ENGINE MISS OR SURGE

NOTE: *Before checking the ignition system, be sure that the fuel system is in good working order.*

Timing, Spark Plugs, Wiring Distributor Cap and Coil

1. Check the timing, wiring, dis-

tributor cap, and the coil tower as you would for a conventional ignition system.

2. Check the spark plugs as you usually would, being sure that they are set to the gap recommended by the manufacturer.

3. Check the ignition coil in the manner outlined in Step 5 of "Engine Will Not Run", above. Replace the coil, if necessary.

NOTE: *The special coil used with this CDI system cannot be tested with a standard coil tester. A normal voltmeter will read zero volts when connected across the ignition coil with the ignition turned on.*

Distributor

The distributor should be tested in the manner outlined in the Distributor Section of "Engine Will Not Run", above.

Amplifier

A poorly grounded amplifier can cause an engine miss or surge. To check, temporarily connect a jumper lead from the amplifier housing to a good ground. If the engine performance improves, the amplifier is poorly grounded. Correct as required.

If no defects up to this point have been found, and the secondary system (plugs, wiring, distributor cap and coil) have been thoroughly checked, the most likely cause of the engine condition is a defective amplifier.

Delco Remy Unit Ignition System

The term "unit" or "unitized" ignition system is used to describe this Delco Remy design because the coil and distributor are both contained in one unit. The distributor is a magnetic pulse type, similar to other Delco Magnetic Pulse distributors, but it has a special distributor cap and a harness assembly molded with all the spark plug wires in one piece. The ignition coil mounts with long bolts on top of the harness assembly and holds the complete sandwich of

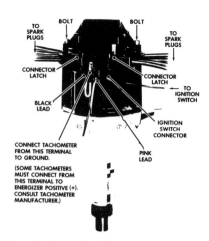

Delco Unitized Ignition. The electronic module plugs coil, harness, and distributor cap together.

TROUBLESHOOTING THE UNIT IGNITION SYSTEM

Electrical problems will usually be found in one of 3 parts: the ignition coil; the pickup coil; or the electronic module, which attaches with screws on the side of the distributor. Testing consists of using an ohmmeter to check the ignition coil and the pickup coil. If they check as good, then the electronic module is probably at fault and will have to be replaced.

To check the coil, connect an ohmmeter as shown in our illustration. Checks A and B should be practically zero. If you get a reading of infinity on either check, the coil should be replaced. Check C should read 6,000 to 9,000 ohms. If the reading is outside this range, replace the coil. Check D should show infinity. If not, the coil should be replaced. To check the pickup coil, attach an ohmmeter as shown in our illustration. Use a hand vacuum pump or a distributor machine to move the vacuum advance throughout the complete range of vacuum, while you are taking the reading with the ohmmeter. Check A should read 650 to 850 ohms. If it reads more or less at any time during the test, replace the pickup coil. Check B is a check for ground. It should read infinity at all times. If not, the pickup coil is grounded and must be replaced.

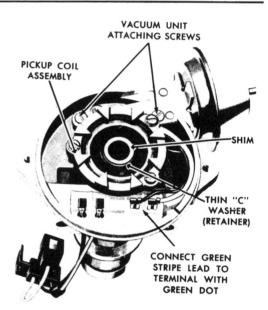

Removing the distributor shaft exposes the pickup coil assembly, with one point for each cylinder

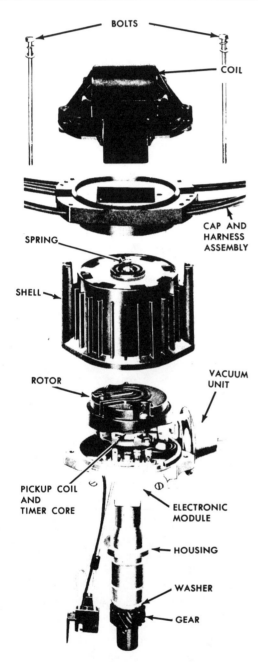

One disadvantage of the Unitized ignition is that the wires are molded into the cap and harness assembly. If one wire goes bad you have to replace all of them

Delco Remy High Energy Ignition System

The high energy ignition system is an improved version of the unitized ignition. It has the advantage of having individually replaceable spark plug wires. The distributor is a magnetic pulse type with an electronic module mounted inside the bowl of the distributor. A special cap is used, with the coil mounted on top of the cap. It is not necessary to remove the coil when removing the distributor cap.

TROUBLESHOOTING THE HIGH ENERGY IGNITION SYSTEM

Ohmmeter checks of the coil should be made as shown in the illustration.

Check No. 1 should be zero or close to zero. If not, replace the coil.

Check No. 2 is to determine if there is an open circuit in the coil wiring. Use the high scale on the ohmmeter and if you get a reading of infinity, the coil is open and must be replaced. If the coil is good, you will get a reading of several 1,000 ohms, but nothing approaching infinity on the high ohmmeter scale. To check the pickup coil for a ground, use an ohmmeter as shown in Step 1. Connect a vacuum pump to the vacuum range, while taking the ohmmeter reading. The reading should be infinity at all times. If not, replace the pickup coil.

Check No. 2 determines the resistance value of the pickup coil itself, which should be 650-850 ohms at all

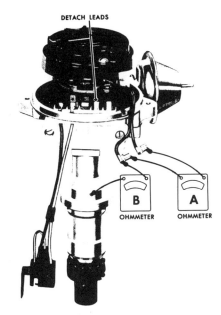

Ohmmeter checks of the pickup coil

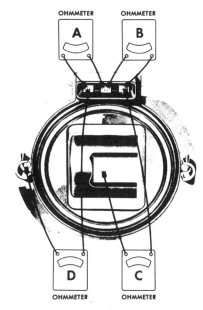

Ohmmeter checks of the ignition coil

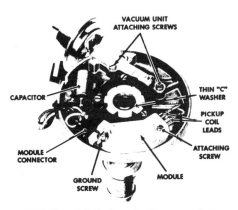

With the distributor shaft removed, the pickup coil and electronic module are exposed

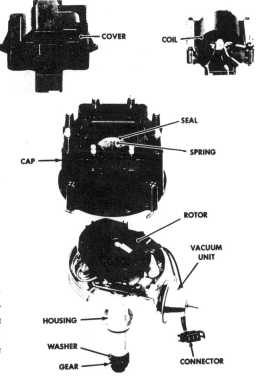

The High Energy Ignition is an improved version of the Unitized Ignition

times. If the pickup coil varies in resistance or is outside those limits, replace it. If the ignition coil and the pickup coil check okay, then the trouble must be in the electronic module and it will have to be replaced.

Ford Motor Company Transistor Ignition

The Ford system appeared about 1962 and was used as original equipment for approximately 5 years. It consists of the conventional distributor with the condenser removed; an amplifier assembly, sometimes called

a "heat sink"; a special high-voltage coil; and a group of 3 units mounted under the hood which are a ballast resistor, a tach block, and a cold start relay. The purpose of the transistor

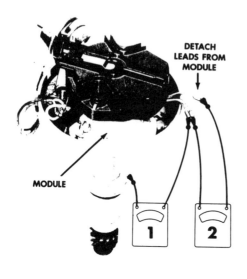

Ohmmeter checks of pickup coil

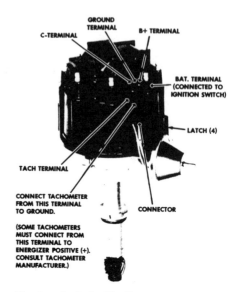

The terminals inside the connector are not shown, but the arrows show their location. Removing the cover exposes the connectors

ignition system, as in all special systems is to give a higher available voltage at high engine rpm and to cut down on the amount of current flow through the points to increase their life. The double ballast resistor in this system is used to protect the amplifier assembly from current surges that might damage it. The tach block is in the system only to provide a place to connect a tachometer. Its terminals are colored red and black for positive and negative tach connections. The cold start relay senses the amount of current the starting motor draws. If the amount of current is excessive so that the cranking voltage drops below $10\frac{1}{2}$ volts, then the cold start relay by-passes the ballast resistor and sends full battery voltage to the ignition coil for a hotter spark during cranking.

Troubleshooting the Ford Transistor Ignition

In case of ignition troubles such as a weak or non-existent spark, connect a dwell meter to the tach block, turn the ignition on, crank the engine, and watch the meter reading. If you get a dwell angle of between zero and 45°, it means that the transistor 's operat-

ing without any trouble. If the dwell meter reads zero, it means the distributor points have high-resistance or are not closing and should be replaced or adjusted. If you get a dwell meter reading of 45°, it means that there is no power from the ignition switch or the distributor points are not opening or the transistor is defective. In this case, disconnect the bullet connector from the distributor lead and crank the engine again with the ignition on and your dwell meter still connected to the tach block. A meter reading of 45° indicates power source or transistor trouble. To find out which, reconnect the bullet connector at the distributor lead and connect a voltmeter to the red terminal of the tach block and to ground, then crank the engine with the ignition switch on. If the voltmeter indicates no voltage, it means that you are not getting any power from the ignition switch. If the voltmeter indicates a steady voltage, the transistor is malfunctioning and the amplifier and transistor assembly will have to be replaced. A voltmeter reading that varies while the engine is cranking would be normal.

The float shuts off the fuel because it has a lot of leverage against the needle. It takes about 8 lbs. of fuel pressure to override a closed needle, much more than a normal fuel pump can put out

When fuel goes out of the bowl through the main jet, the float drops, and opens the needle to let more fuel in. Vertical needles are the best design because they are self-cleaning. Dirt falls past the needle into the bowl instead of sticking in the seat

How Carburetors Work

Carburetors vary greatly in the way that they are made, but they are all basically the same in that their job is to mix the air with the fuel in the proportion that the engine needs. There are six different systems or fuel/air circuits in a carburetor that make it work. These systems are the Float System; Main Metering System; Idle and Low-Speed System; Accelerator Pump System; Power System; Idle and Low-Speed System; and the Choke System. The way these systems are arranged in the carburetor determines what the carburetor looks like.

It's hard to believe that the little single-barrel carburetor used on 6 cylinder or 4 cylinder engines have all of the same basic systems as the enormous 4-barrel carburetors used on the big V8s. Of course, the 4-barrels have secondary throttle bores and a lot of other paraphernalia that you won't find on a single-barrel carburetor. But basically, all carburetors are the same and if you can understand a simple single-barrel, you can use that knowledge to understand the 4-barrel. Understanding the basic systems in a carburetor is important from the standpoint of diagnosis. If you have a lean-running engine, you must have enough knowledge about carburetors to know that it might be caused by a low float level or the wrong jet size. The more you know about carburetors, the easier it will be to make this kind of quick decision that will save you time when you are trying to find a problem.

FLOAT SYSTEM

When the fuel pump pushes fuel into the carburetor, it flows through a seat and past a needle which is a kind of shutoff valve. The fuel flows into the float bowl and raises a hinged float so that the float arm pushes the needle into the seat and shuts off the fuel. Floats have a large range of movement when you look at them in a disassembled carburetor, but in actual operation they move very little, as long as the fuel pump has enough capacity to keep up with the engine. If the engine uses more fuel than the pump can supply, then the float drops down, which allows the fuel to push the needle valve away from the seat and more fuel flows in. If the float system is properly designed, the fuel level will stay fairly constant which means that the engine gets the right mixture at all times. Floats are adjusted by bending a little tang on the float arm that bears against the needle.

All manufacturers give a float level specification. If the float is the type that is mounted in the bowl itself, the float level specification is a measurement from the top of the bowl to the top of the float. If the float is the type that is mounted on the bowl cover, then the measurement is taken from the cover to the top of the float. The float level measurement is always taken with the needle in the closed position and the float resting against it. If the float is in the bowl you can close it by filling the bowl with solvent or by turning the carburetor upside down. Some manufacturers recommend pushing on the little float tang to hold the float and needle in the closed position. We do not recommend this because it requires a delicate touch so that you don't push the soft rubber tip of the needle into the seat and make a groove in it. If the float is attached to the cover, then the way to close the needle is to turn the cover upside down, so that the float bears on the needle. Then make your measurement between the float and the underside of the bowl cover.

Fuel level is also given as a specification for some carburetors. The fuel level is the measurement from the top of the bowl to the surface of the fuel in the bowl. Of course, measuring the actual wet fuel level is more accurate than just measuring the float level, but the measurement of fuel level cannot be done on all carburetors. On some carburetors, there are holes in the top of the bowl that can be uncovered by removing parts of the carburetor, which will then allow you to stick a scale down through the hole and measure the actual wet fuel level. Other carburetors have sight holes covered up by plugs in the side of the bowl. When these plugs are removed

with the engine idling, the fuel should just wet the threads in the bottom of the hole. If the fuel runs out, the fuel level is too high.

In some manuals and even in some shops, you may see a special gauge that is attached to the carburetor, so that the fuel in the bowl can run out into the gauge and tell you the actual wet fuel level. Whichever you use, fuel level or float level, you should make adjustments to the float according to the manufacturer's specifications. In most cases, the float level is set and the fuel level is an optional check that you can use if you feel it is necessary.

Float drop is the measurement of the amount that the float drops when there isn't any fuel in the bowl. Float drop is regulated by a second tang on the float arm which usually hits against the side of the brass seat. Float drop is not a precise measurement. As long as the float drops enough to allow the needle to open and fuel to enter, it will probably be alright. However, you must be sure that the float cannot drop so far that the needle jams against the float tang. Some carburetors have a boss in the bottom of the bowl that keeps the float from dropping too far. In those instances, there is no float drop adjustment because the boss takes care of it. In any case, follow the manufacturer's instructions. If he gives a float drop specification, then by all means check it when you have the carburetor apart.

A lot of development has gone into

The venturi causes a vacuum at the end of the main nozzle, and fuel is pulled out of the bowl

The vacuum in the venturi is dependent on the flow of air. With no airflow, there is no vacuum

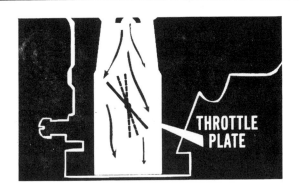

The throttle plate regulates airflow into the engine, but more important, it controls the density of the mixture. At the high vacuum of a closed throttle, the mixture is thin, and very little power is produced. At wide open throttle there is more fuel going into the engine, and more power is produced

THROTTLE PLATE

the needle and seat in most modern carburetors. Originally, needles were all steel and seats were all brass. The trouble with a steel needle on a brass seat is that it can be held open by a very small amount of dirt. The manufacturers have gone to great lengths to try to solve this problem, even using diaphragm needles that were very difficult to repair. Today, almost every carburetor uses a steel needle with a rubber or Viton® tip, or a complete steel needle that bears against a rubber or Viton® seat. This flexible needle tip or seat is the reason that you must be careful not to push too hard against the seat when checking float level. The needle can get a groove in it that will take several hours, if ever, to go back to its original shape.

MAIN METERING SYSTEM

At cruising speed, most of the fuel for engine operation comes through the main metering system. When the engine is running, it constantly sucks air through the throat of the carburetor. The throat is made with a narrow portion in the middle, called the venturi. When the air passes through the venturi section of the throat, it creates a slight vacuum in that area. The vacuum acts on the end of the main nozzle, which is a tube that runs at a slant from the bottom of the float bowl up into the venturi. The slant is necessary so that the fuel will not run out of the float bowl into the throat of the carburetor. Some carburetors use multiple venturis, with one inside the other and the end of the main nozzle

inside the smallest venturi. These multiple venturis increase the suction effect on the end of the main nozzle.

Fuel would flow from the float bowl through the main nozzle in tremendous quantity if it were not for some type of metering. The fuel is metered by a calibrated hole called the main jet, located at the bottom of the float bowl. All of the fuel has to pass through the main jet before it can get to the main nozzle. In some carburetors, the jet is connected directly to the end of the main nozzle, but in most modern carburetors there is a main well between the jet and the bottom end of the nozzle. This main well contains screens or bars to break up the fuel and also is the area where air bleeds allow air to mix with the fuel and try to break it up before it goes out the end of the main nozzle. Raw fuel does not travel too well through an intake manifold. It has a tendency to fall out of the air stream. So everything in the carburetor is designed with the idea that the finer that you can break up the particles of fuel, the better the engine will run and the better will be the fuel distribution.

Some carburetor designs use a main jet and let it go at that. Others have a main jet with a metering rod resting in the jet. The metering rod is simply a brass rod that hangs in the jet hole and limits the amount of fuel that can go through. Metering rods have tapers or steps on them. The rod gets thinner toward the bottom end. If the rod is lifted slightly, so that the center portion is in the jet hole, then more fuel will flow and the mixture will become richer. This richer mixture is needed as the engine goes faster. There are several ways to lift the metering rod in the jet. It can be done mechanically, through a connection with the throttle linkage or it can be done by vacuum, acting on a piston or a diaphragm. The vacuum arrangement is setup so that it holds the metering rod down in the jet. As the engine throttle is opened and more load is put on the engine, the vacuum reduces and the metering rod slowly rises up out of the jet from the force of a spring underneath the vacuum piston. Because the suction in the venturi that draws the fuel out of the end of the main nozzle is so small, the fuel level in the bowl is critical. The main nozzle slants upward with its venturi end higher than the fuel in the bowl, so that the fuel level in the bowl is also the level of fuel inside the main nozzle. Ideally, the fuel should stay just inside the tip of the nozzle, ready for the slight venturi vacuum to pull it out, but not dripping. If you have ever seen a car-

buretor that drips when the engine is off, it means that the fuel level in the bowl is too high. If the fuel level is too low, the fuel will be way down inside the main nozzle and the slight venturi vacuum will have difficulty pulling it out. This causes a lean mixture and makes a poor running car at cruising speed.

What we have been describing so far is the main metering system as it exists in a single throat carburetor.

The main metering system works very well at cruising speed, but it depends on the airflow in the venturi to have enough suction to draw the fuel out of the main nozzle. At very low speeds and at idle, the main metering system simply won't work. In order to have the car run at slow speeds, we have to have the Idle and Low Speed System.

IDLE AND LOW-SPEED SYSTEM

The vacuum in the intake manifold at idle is high because the throttle is

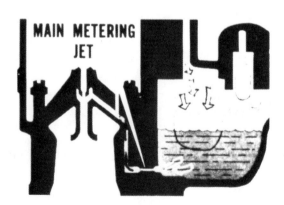

Fuel is drawn out of the float bowl through the main jet, and up into the main nozzle. Actually, atmospheric pressure entering the bowl vent pushes the fuel out

As the fuel travels between the main jet and the main nozzle, it picks up air from air bleeds. The air helps to vaporize the fuel and get better distribution in the manifold

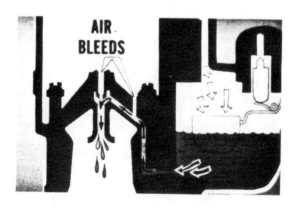

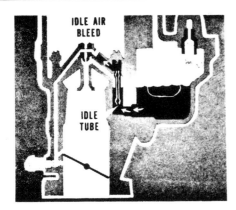

The idle tube is a calibrated jet that picks up fuel from the main well or the float bowl. The idle passages go up, over, and down to the idle port under the throttle blade

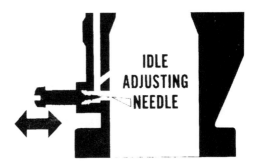

When the mixture from the idle system mixes with the air entering the engine, it forms the fuel-air mixture that the engine uses to idle. The hole above the needle is the idle transfer port, which feeds more fuel as the throttle opens, for a smooth transition from idle to cruising

almost completely closed. This vacuum is used to draw fuel into the engine through the idle system and keep it running. The idle jet is usually a tube with a calibrated hole in the end, that sticks down from the top of the carburetor in tne main well, below the fuel level. The upper end of the tube connects to a passageway above the fuel level, which crosses over and then travels down inside the casting, to a hole in the throat of the carburetor below the throttle valve. Engine vacuum acts on the hole and sucks the fuel up into the idle jet, across, and down into the engine. As the fuel crosses over above the fuel level, it is mixed with air from air bleeds in the top of the carburetor. There may be some restrictions in the channel that help to break up the fuel and mix it with the air.

The fuel cannot be allowed to run uncontrolled into the engine, through the hole under the throttle valve, so an idle mixture needle is built into the carburetor with its point resting in the hole. When the needle is screwed in, it closes the hole off partway and limits the amount of fuel that can go into the engine. When the mixture is properly adjusted with the needle, it mixes with the air passing around the throttle blade and gives the engine the right fuel and air mixture for a good idle. There is usually one idle mixture needle for each throat in the carburetor,

but there have been designs in the past that use a single mixture needle located high up on the carburetor that controls the mixture to more than one throat at a time.

Other designs of carburetors close the throttle valve completely at idle and use a large screw to allow the air to by-pass the throttle valve. Instead of opening the throttle to adjust engine speed at idle, you adjust the by-pass air screw.

As the throttle opens, the venturi main metering system does not take over immediately. The throttle has to be open fairly far before there is enough airflow through the venturi to suck fuel out of the main nozzle. To take care of this transistion period and keep the engine running smoothly, there is what is known as an off idle port or an idle transfer port above the throttle valve. As the throttle valve opens, it exposes this port to engine vacuum and we have an additional flow of idle mixture.

The idle and low-speed system is commonly thought of as working only during idle, but actually it feeds fuel almost up to wide-open throttle. It's important to remember that the idle mixture needle does not change the mixture in the idle passageway from rich to lean. It only changes the *amount* of mixture coming through the passageway. The richness or leanness of the idle passageway mixture is controlled by the size of the idle

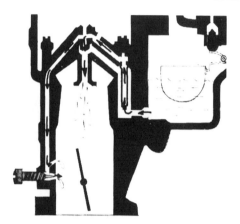

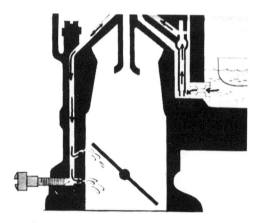

Both the idle mixture port and the trans-fer port are feeding fuel into the throat as the throttle is opened

As the throttle opens wider, the main system starts to feed fuel, but the idle system still feeds fuel also, although not as much

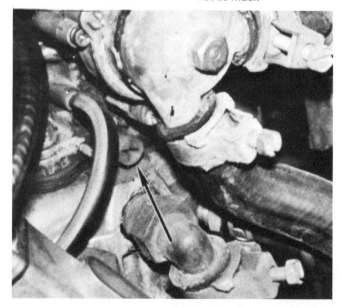

Idle mixture screws can be hard to find, as on this Pinto, where the screw is hidden under the choke housing

When the throttle moves from the open to the closed position, the accelerating pump draws in fuel from the float bowl, through the inlet check

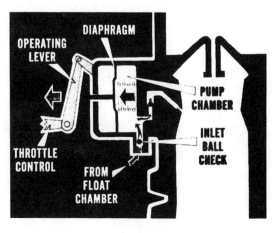

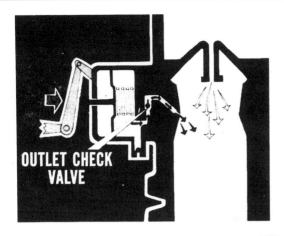

When the throttle is opened, the pump forces the fuel through the outlet check and into the carburetor throat. The inlet check closes to prevent fuel flow back into the bowl

When the pump fills, the outlet check closes to prevent air from being sucked into the pump chamber

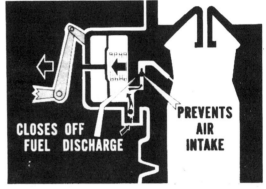

jet, which in many carburetors is not replaceable. If the idle jet is the wrong size or damaged so that it limits the flow of fuel, you could compensate for this at idle by unscrewing the idle mixture needle and allowing more mixture to mix with the air coming around the throttle valve. However, as the throttle opens and exposes the idle transfer port, the restrictive idle jet would cause a lean mixture in the transition period, contributing to a stumble. It works the other way around too. If the idle jet is too big it can cause a rich mixture throughout the cruising range, almost up to top speed.

ACCELERATOR PUMP SYSTEM

When the throttle is opened, the air flowing through the venturi starts moving faster almost instantly, but there is a lag in the flow of fuel out of the main nozzle. The result is that the engine runs lean and stumbles. It needs an extra shot of fuel just when the throttle is opened. This extra shot of fuel is provided by the accelerator pump. It's nothing more than a little pump operated by the throttle linkage that shoots a squirt of fuel through a separate nozzle into the throat of the carburetor. The accelerator pump jet, which is usually located in the pump nozzle, is calibrated so that it supplies the right amount of fuel.

The pump is usually a plunger working in a vertical cylinder. As the throttle is closed, the plunger rises and draws fuel into the cylinder, from the float bowl, which stays there ready for the next time the throttle is opened. When the throttle is opened the plunger is forced down the cylinder, which pushes the fuel up through the nozzle into the throat of the carburetor. When the pump is on the upstroke drawing fuel in from the float bowl, an intake check, usually a steel ball, opens to allow the fuel to enter from the bowl. An outlet check closes at the same time so that the pump will not suck air back from the throat of the carburetor. When the pump moves down to push the fuel into the throat, it opens the outlet check and closes the intake check so that fuel will not be forced back into the carburetor bowl.

The intake check is not on all pumps. Some allow the sides of the pump itself to collapse on the upstroke and the fuel flows past the lip of the cup, into the bottom of the pump well. A vertical plunger pump design can be operated directly by the throttle linkage or operated by a spring around the pump shaft. When the pump is operated by the spring, the throttle linkage holds the spring in a compressed position. When the throttle is opened, the linkage relaxes its hold on the spring and the spring pushes the pump plunger down. The spring gives the same rate of flow everytime the pump is operated and prevents an overly ambitious driver from bending the linkage by shoving the throttle down too fast.

Pumps can also be a rubber or neoprene diaphragm connected to the throttle linkage. The diaphragm has the disadvantage that it has to be placed below the fuel level, in order to receive fuel and stay full at all times. If the diaphragm develops a leak, as it will when it eventually wears out, fuel will leak out of the carburetor onto the intake manifold.

Most pumps can be adjusted for the length of the stroke with different holes in the pump link. If there are no optional holes, the linkage itself can be bent to change the stroke of the pump. A shorter stroke results in less fuel being squirted into the carburetor throat and a longer stroke more fuel. Flat spots in acceleration can be caused by too much accelerator pump fuel the same way they can be caused by too little. In the days before emission controls, such adjustments could be easily made to tailor the pump setting to the car so that all flat spots could be eliminated. In these days of emission controls, pump adjustments must be made according to the manufacturer's specifications. Unauthorized adjustments may change emission levels and should not be done unless the adjustment is factory authorized.

POWER SYSTEM

The main metering system works very well at normal engine loads, but when the throttle is in the wide-open position, the engine needs more fuel to prevent detonation and give it full power. The power system provides additional fuel by opening up another passageway that leads to the main nozzle. On some carburetors, this extra passageway is controlled by a power valve. The power valve can be opened by a vacuum-operated metal

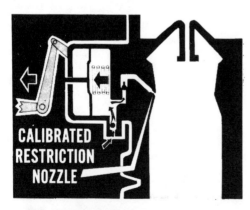

The accelerating pump jet is usually visible in the throat of the carburetor. On many carburetors it is not replaceable. To change the size it must be drilled larger, or soldered up and drilled smaller

In a piston actuated power valve, manifold vacuum holds the piston up. When the throttle is opened, the piston drops and opens the valve

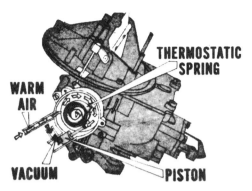

Older chokes with a built-in vacuum piston use vacuum to pull the choke open, and to draw hot air into the choke housing

The choke plate not only cuts down on the amount of air entering the engine, but causes more fuel to flow from the jets because of the increased vacuum below the choke plate

Most modern chokes are the well type, which use direct heat from the manifold to heat up the choke coil. Some also use an electric heater, as shown here

plunger, a diaphragm, or by mechanical linkage from the throttle. If the power valve is vacuum-operated, the high vacuum in the engine keeps the valve closed above approximately 10 in. Hg. When the throttle is opened and the vacuum drops, the spring behind the plunger or diaphragm pushes the power valve open and allows the additional fuel to flow up to the main nozzle. If the power valve is operated by mechanical linkage from the throttle, then it is strictly a question of throttle position and has nothing to do with the actual load on the engine. Most of the power valves today are vacuum-operated.

Some carburetors do not use a separate power valve. They consider that lifting the metering rod or what they may call a "step-up" rod out of the jet is adequate without the need for a

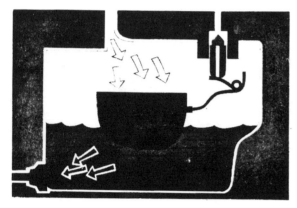

A bowl vent is necessary for two reasons. Air must be able to go out of the bowl when fuel enters, and come into the bowl when the fuel leaves

separate power valve. The lifting of the metering rod or step-up rod out of the jet is accomplished by a spring underneath a vacuum piston. When engine vacuum falls off, the spring pushes the rod up and the thinnest portion of the metering rod or step-up rod is in the jet, which allows more fuel to flow.

CHOKE SYSTEM

A choke is necessary on a cold engine because the fuel condenses out of the air/fuel mixture on to the combustion chamber and cylinder walls. This means there is less fuel in the mixture to actually burn and some means must be used to get more fuel into the engine. The choke does this because it is a plate located at the top of the carburetor, above the venturi. When the choke is closed, very little air can enter the engine. This in itself would richen the mixture, but the main reason that the mixture is richened when the choke closes is that the entire vacuum suction from the engine acts on the main nozzle and literally sucks raw fuel right out of the carburetor bowl. Of course, we very seldom run an engine with a completely closed choke, but moving your choke just a few degrees toward the closed position will create a vacuum underneath it that will draw more fuel out of the main nozzle.

Chokes used to be manually controlled by the driver, which worked fine as long as he remembered to shut it off after the engine warmed up. Now most chokes are automatically

controlled by a thermostatic spring, either on the carburetor or mounted on the engine and connected to the carburetor through a linkage. When the thermostatic spring is cold, it pushes the choke towards the closed position. If the spring is mounted on the carburetor, hot air from around the exhaust manifold is piped to the spring to make it open quickly. If the thermostatic spring is mounted in a well on the intake manifold or next to the intake manifold, it picks up heat from the exhaust crossover passage.

When a cold engine is being started, we need a completely closed choke for quick starting, but the instant the engine starts running, we need to get that choke open to give the engine enough air to keep going. The thermostatic spring does not heat quickly enough and open the choke as fast as needed, so that a piston or diaphragm, called a vacuum break diaphragm, is used to pull the choke open just a few degrees; enough so that the engine will keep running. The piston type of vacuum break use to be built into the choke housing and it was common to have heat distort the housing and the piston to stick. Today, the vacuum break is a separate diaphragm which is much more reliable than the older piston.

Another device to keep the engine running after it starts is the fast idle. When the choke is closed, linkage connects to a fast idle cam and raises it underneath the throttle stop screw, so that the engine idles at a faster speed than normal. This fast idle cam

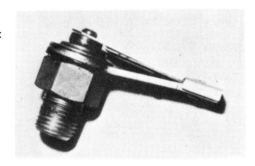

This hot idle compensator is a screw-in type that might be found off or on the carburetor. This particular one was mounted in the PCV vacuum line on a Ford

must be set by depressing the throttle before starting a cold engine. In fact, the fast idle cam is connected to the choke plate itself so that if the fast idle cam is not set, the choke will not close.

The type of choke that mounts on the carburetor itself usually has a housing with marks on it and can be adjusted by lining up the marks according to factory specifications.

The type of choke coil that mounts off the carburetor is usually adjusted, if necessary, by bending a link connecting the choke coil to the choke plate. A close inspection of the choke coil that mounts off the carburetor will reveal marks on it and an adjustment nut on some models. However, this adjustment is usually for factory adjustment only. The way to adjust the choke in the field on this type is by bending the link. Other manufacturers, who have their chokes mounted off the carburetor, do not provide any adjustment specifications and they do not recommend that you reset the marks or bend the link. In all choke adjustments, you must follow the manufacturer's specifications.

Choke Unloader

If the driver accidentally floods the engine by working the throttle too much and squirting too much fuel out of the accelerator pump, the engine won't start unless the choke can be opened. The unloader device takes care of this by providing a choke opening of a few degrees, just enough to get the engine started if the driver will shove the throttle all the way to the floor and hold it there. Unfortunately, many drivers don't know

about the unloader feature and therefore the flooded carburetor still presents quite a few problems on a cold engine.

OTHER COMPONENTS

Bowl Vents

If you want to be technical about it, the suction in the venturi really does not draw the fuel out of the main nozzle. What happens is that atmospheric pressure, acting on the fuel in the bowl, pushes the fuel out of the main nozzle because there is less pressure at the end of the nozzle. We mention this because it will help to understand why bowl vents are necessary. If the atmospheric pressure can't get into the bowl to act on the fuel, then it can't push it out of the nozzle and the engine won't run, or at least it won't run very well. Bowl vents are sometimes necessary when an engine stops in order to get rid of evaporating fuel that might go down into the throat of the carburetor and make the engine hard to start.

When vapor emission controls came out, bowl vents on many carburetors were either eliminated or the vent was connected to the carbon canister so that the vapors could be collected. Some carburetors still use a bowl vent that goes into the atmosphere but it has a valve in it that only opens when there is high pressure in the bowl. This evidently doesn't happen very often, so that the system is considered non-polluting.

Bowl vents on some carburetors are what is known as an internal vent. The air space over the fuel in the bowl is connected by a slanting tube into the throat of the carburetor above the choke valve. This internal vent not only keeps the fuel vapors inside the air cleaner, but helps the fuel flow at high rpm by allowing the rush of air through the throat of the carburetor to go down the tube and help push the fuel up into the main nozzle.

Carburetor Heat

The mixture passages in the intake manifold under the carburetor have been heated for many years in both inline and V8 engines by exhaust passages that run through the manifold. This intake manifold heat is necessary to eliminate flat spots and get better fuel distribution because there is more complete vaporization of the fuel. Fuel does not vaporize as easily as you might think, and the heavy particles of fuel that tend to settle out of the air moving in the intake manifold can make some cylinders run rich, while a cylinder right next to it is running lean. Actually, carburetor heat, in part, makes up for deficiences in intake manifold design. For many years, a heat valve was used in the exhaust manifold on both inline and V8 engines to send some of the exhaust through the heat passages in the intake manifold. The valve was spring-loaded with a thermostatically-controlled spring, so that at high engine temperatures it did not send as much exhaust through the heat passages. Now most engines use a heated air cleaner which sends hot air to the carburetor on a cold engine and shortens the warm-up time considerably. Many manufacturers have found that they can do away with the heat valve in the exhaust manifold when they have the heated air cleaner. In most cases, the heat passages are still in the intake manifold, but the valve has been eliminated from the exhaust manifold.

Hot Idle Compensators

When underhood temperatures get extremely high on a hot day, the fuel vaporizes much easier. This extra vaporization of the fuel can cause a rich mixture, particularly at idle. The hot idle compensator is a little valve which can be mounted on the carburetor or in a vacuum hose, such as the PCV hose. When underhood temperatures get extremely hot, the hot idle compensator valve opens and allows atmospheric air to flow into the intake manifold. This leans the excessively rich mixture down to where the engine will keep idling without stalling. The hot idle compensator opens at idle because that is the period when there is no airflow (or very little airflow) through the engine compartment and very high temperatures result. When the car starts moving, the airflow increases and that usually is enough to cool the compensator and close it.

Vacuum Control Ports

The carburetor not only has the job of mixing the air and the fuel, but also of regulating vacuum supply to various other parts of the engine, such as the distributor vacuum advance and emission controls.

PORTED VACUUM

Ported vacuum is a system of vacuum advance control that results in no spark advance at idle. A vacuum port is located in the throat of the carburetor, so that it is above the throttle valve when the throttle is in the curb idle position. There is no vacuum above the throttle valve, so that the distributor vacuum advance goes to the neutral or no advance position at idle. When the throttle is opened above idle, the vacuum under the throttle blade acts on the spark port and advances the spark. The ported vacuum system is also used to operate exhaust gas recirculation valves. Sometimes, the EGR valve is operated from the same port that operates distributor advance. On other carburetors, there may be two separate ports. There are even emission control systems where the operation of the distributor advance is switched back and forth from the EGR port to the spark port.

Another system operated by a port over the throttle blade is the purge to the vapor control canister. Purge at idle might upset the engine idle, therefore the port is placed above the throttle blade and the canister is only purged when the throttle is opened.

VENTURI VACUUM

Venturi vacuum is another control method that was formerly used to operate the Ford all-vacuum distributor several years ago. Venturi vacuum works from a port in the venturi of the carburetor. The low pressure or suction in the venturi area acts on the port the same way it does on the main nozzle. Venturi vacuum is not very strong, but when used with a distributor vacuum advance that is calibrated for it, it will operate the vacuum advance according to the amount of airflow that is rushing through the carburetor throat. This airflow is inversely proportional to the load on the engine. When the load is high, the throttle is wide-open. Therefore, the airflow is low, which means that venturi vacuum is low and the spark is retarded. When the load is light, the airflow is high and the spark advances.

On some emission controlled cars, the venturi vacuum system is used to operate an exhaust gas recirculation amplifier. The venturi vacuum turns on the amplifier valve and then the amplifier valve sends full manifold vacuum to open the EGR valve. Because there is no venturi vacuum at idle, the EGR valve stays closed. There is venturi vacuum at wide-open throttle, but manifold vacuum is so low that there isn't enough to open the EGR valve. Venturi vacuum is not necessarily any better or any worse than the vacuum system. It's just another way of trying to get precise control over the vacuum-operated units.

Anti-Stall Dashpots

When the driver of a car with an automatic transmission suddenly takes his foot off of the throttle, the throttle closes before the engine gets a chance to slow down. The engine is turning faster than it would at a normal curb idle, but the carburetor having its throttle in the idle position, is supplying a curb idle mixture. The result is the engine won't run and it stalls. An anti-stall dashpot, which is sometimes called a slow closing throttle, is used on the throttle linkage so that the throttle closes slowly whenever the driver lifts his foot. This gives the mixture a chance to stabilize and the engine keeps running. You will also find these dashpots on some manual transmission cars. In some cases, they have been used to prevent driveline whip which is damaging to the manual transmission driveline. In other cases, they have been used as an emission control device to prevent the excessively rich idle mixture from being pulled through the engine before it gets a chance to slow down to idle speed. When you see an anti-stall dashpot on a manual transmission car, the chances are that it was not a mistake, but it was put there for a definite purpose.

Anti-Dieseling Solenoids

When the emission controls came out, the manufacturers generally went to retarded spark and greater throttle openings at idle. This gave a better mixture that burned more completely for fewer emissions, but it created a problem of dieseling or after-running; i.e., the engine would keep going after the ignition switch was turned off. Lean mixtures, high combustion chamber temperatures and the increased throttle opening at idle all contributed to the dieseling problem. One way to stop dieseling is to shut off the air, or at least cut down on the air, that the engine receives. The anti-dieseling solenoid does this by providing one throttle setting for normal curb idle and another throttle setting which is considerably smaller. When the anti-dieseling solenoid is used, the normal curb idle is set with the solenoid adjustment. It can be a threaded stem, a solenoid that moves in a bracket, or is adjusted in some way to give the correct engine idle speed. The solenoid is connected so that it is on whenever the ignition switch is on. When the driver turns the ignition off, the anti-dieseling solenoid also goes off, allowing the throttle to drop that back to a much smaller opening. This cuts down on the air entering the engine and supposedly eliminates dieseling.

The manufacturers give idle speed specifications for anti-dieseling solenoid equipped cars as two numbers separated by a slash. The lower number is the speed that is set with the solenoid off. This adjustment is made with the normal throttle stop screw. When the solenoid is on, the stem of the solenoid holds the throttle open off the normal stop screw and the adjustment for curb idle is made with the solenoid itself.

Anti-dieseling solenoids are strictly an electrical unit. If you see a solenoid that looks like an anti-dieseling solenoid, but has vacuum hoses connected to it; it is there for another purpose entirely. The solenoid with vacuum hoses to it is the General Motors CEC valve, which is not for anti-dieseling, but for holding the throttle open during deceleration.

Throttle Openers

For several years General Motors has used a throttle opener called a CEC valve. The CEC valve is a Combined Emission Control that regulates the transmission controlled spark and also opened the throttle for a wider closed throttle opening during deceleration in High gear. Idle speed should never be set with the stem of the CEC valve. The valve can be recognized because there are vacuum hoses running to it. There is a specification of engine rpm for the CEC valve, but it has nothing to do with idle speed and it's only necessary to set it after carburetor overhaul or any other reason for disassembling the throttle linkage.

Imported cars also use deceleration throttle openers of varying designs. The throttle opener is used for only one reason, to prevent the excessively rich mixtures that come from deceleration in High gear or at high speeds with a closed throttle. Imported car throttle openers are usually 100% vacuum-controlled although some of

them are connected with a speed switch that shuts off the throttle opening below a certain speed.

If you drive a car with a throttle opener it feels as if it's trying to keep going when you take your foot off the gas in High gear. This is not a defect, it's designed that way deliberately. If the car feels like it is running away, it's possible that the throttle opener is set too high.

Deceleration Enrichers

Small engines have a problem with maintaining combustion when they decelerate under closed throttle. Many 4 cylinder engines use what is called a "deceleration enriching valve" or a "coasting enricher." In the case of the Pinto and the Capri, this valve is separate, mounted on the intake manifold. It sucks fuel and air out of the float bowl of the carburetor and allows it to enter the intake manifold to support combustion during deceleration. Imported cars use a similar system, but the deceleration enricher is built into the carburetor itself. It's similar to a diaphragm-operated power valve, but it only opens during deceleration.

Secondary Carburetor Throats

At low engine speeds, a car runs best with a small carburetor and at high engine speeds, it puts out its best power with a large carburetor. The problem is to build both carburetors into the same engine. The solution is a system of primary and secondary throats. The primary throttles work in all normal, around town and low-speed driving. When the primary throttles are opened to about a 3/4 opening, then the secondary throttles start to open and both primary and secondary throttles reach the wide-open position at the same time. The result is that the engine runs well at low-speed and puts out maximum power at high-speed. Most secondary throats have only a main metering system, but some of them may have a constant flow idle system that makes the engine idle smoother.

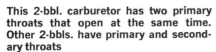

This 2-bbl. carburetor has two primary throats that open at the same time. Other 2-bbls. have primary and secondary throats

In normal cruising a 4-bbl. carburetor runs on the two primary barrels

On domestic vehicles, a two-barrel carburetor has always meant a carburetor with two primary throats. However, imported cars have two-barrel carburetors on their small 4 cylinder engines that have primary and secondary throats.

Four-barrel carburetors are all of the primary/secondary design with two primary throats that are for normal running and two secondary throats that open at high-speed or high-load. The activating control to the secondary throttles can be either mechanical through throttle linkage, or through a venturi vacuum system. If a large enough diaphragm is used, the venturi vacuum is strong enough to actually open the throttle plates at high airflow.

Vacuum-operated secondaries only open when the engine needs them. It is impossible for the driver to get them open at low speeds. Mechanical-ly-operated secondaries are different. If the driver doesn't know what he is doing, he can stick his foot all the way to the floor, opening all 4 barrels wide-open when the car is only going two mph. This can cause the engine to gasp and take several seconds to recover. To keep this from happening, some carburetors have a weighted or spring-loaded air valve in the secondary barrels, above the throttle plates. The air valve opens only if the airflow through the secondary throats is high enough to push it open.

Another problem with the driver, is that he sometimes insists on opening all 4 barrels when the engine is cold. Calling on a cold engine to produce maximum horsepower can be very damaging. Carburetors with secondaries get around this problem by using what is known as a "lock-out". A little lever connected to the choke

In high speed driving all four barrels open up. On a carburetor with mechanically controlled secondaries, flooring the throttle at low speeds will also cause all four barrels to open up

drops down over the secondary and prevents their opening until the choke is fully opened. The lock-out on some 4-barrel carburetors is not on the throttle plates themselves, but on the air valve above the throttle plates. The throttle plates open anytime the driver floors the gas pedal, but it doesn't do anything unless the choke is off because the little lock-out lever keeps the air valve closed.

Vacuum-operated secondaries don't need a lock-out mechanism because they would not open at low-speed anyway, as they depend upon airflow through the venturi to get enough vacuum to open the secondaries.

Component Identification

PCV (POSITIVE CRANKCASE VENTILATION) VALVE

The PCV valve is used to meter the flow of air through the positive crankcase ventilation system. It is also used to prevent fuel vapors in the crankcase from being ignited in case of a backfire.

On six-cylinder engines, it is usually mounted in a grommet on the valve cover. On V8 engines it may be located on the intake manifold, the valve cover, or in the middle of the PCV line, itself.

AIR INJECTION SYSTEM

Antibackfire (Diverter) Valve

The antibackfire (diverter) valve is used to prevent backfiring during deceleration by venting the pump output into the atmosphere.

Several different valve designs are used, depending upon manufacturer. The valve may be located in the line which runs from the pump to the injection manifold or on the back of the pump.

Typical PCV valve

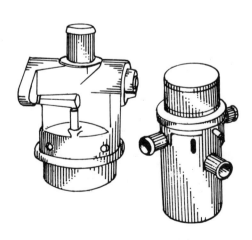

Anti-backfire (diverter) valve

Anti-backfire (diverter) valve

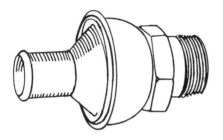

Check valve

Check Valve

The check valve is used to prevent hot exhaust gases from flowing back into the air injection system, in case the air flow should cease, due to a broken drivebelt or seized pump.

The check valve is either located in the line which runs between the anti-backfire valve and the air injection manifold or on the air injection manifold itself.

DISTRIBUTOR CONTROLS

Timing Retard Solenoid

The timing *retard* solenoid is used on some Chrysler Corp. cars up to, and including, 1971. It is used to re-

Timing retard solenoid
(© Chrysler Corp)

duce engine emissions during idle and deceleration.

The solenoid is mounted on the distributor vacuum unit and cannot be serviced separately from it.

Timing advance solenoid
(© Chrysler Corp)

Timing Advance Solenoid

The timing *advance* solenoid is used on some Chrysler Corp. products starting in 1972. It is very important in servicing this item that it is not confused with the timing retard solenoid above.

The timing advance solenoid is used to provide easier engine starts by advancing the timing during cranking.

The advance solenoid, like the retard solenoid, is mounted on the distributor vacuum unit and cannot be serviced separately from it.

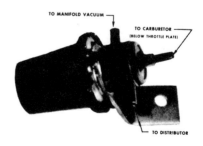

Deceleration valve
(© G.M. Corp)

Deceleration Valve

The deceleration valve is used to provide maximum vacuum advance during engine deceleration. This causes more complete combustion of the fuel/air mixture.

The deceleration valve is located in the vacuum line which runs to the distributor vacuum advance unit.

Typical coolant temperature operated vacuum valve
(© Ford Motor Co)

Coolant Temperature Operated Vacuum Valve

NOTE: The coolant temperature operated vacuum valve is also referred to as a thermal vacuum valve, coolant temperature override switch, thermal vacuum switch, or ported vacuum switch.

The coolant temperature operated vacuum valve is used to provide intake manifold vacuum to the distributor vacuum unit, if the engine overheats during prolonged periods of

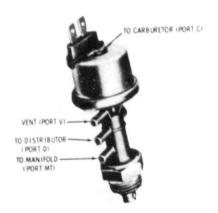

Oldsmobile combination vacuum control switch
(© G.M. Corp)

idling; this advances the timing which causes an increase in engine speed. When the coolant temperature returns to normal, vacuum is blocked and the engine slows down to its regular idle speed.

On cars which are also equipped with a distributor vacuum control system, the valve is used to override the system under overheating conditions. On some engines it is used to override the EGR system.

The valve is threaded into the water jacket and is usually located toward the front of the engine.

Oldsmobile Combination Vacuum Control Switch

Some 1971-72 Oldsmobiles and some 1972 Buicks use a coolant temperature operated vacuum valve which is combined with a solenoid. The solenoid is operated by a transmission switch and is designed to provide vacuum advance only when Third or Fourth gear is engaged.

Vacuum advance is also provided when coolant temperature goes above 218-224°F, to allow an increase in engine speed, which will cause the engine coolant to return to normal temperature.

The combination vacuum control switch is threaded into the coolant passages at the front of the block.

Spark Delay Valves

There are several types of spark delay valves in general use. Some merely retard the vacuum to the distributor vacuum unit so that vacuum advance is delayed for a predetermined length of time.

Some other types of spark delay valves function only when the underhood temperature is above a specified level. Until the temperature reaches this level, normal vacuum is provided to the distributor vacuum unit; above this, the vacuum signal is delayed.

Spark delay valves are located in the line that runs from the vacuum source to the distributor vacuum unit. Be careful not to confuse the spark delay valve with other components of the emission control system which may resemble it.

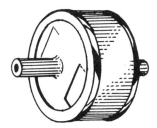 

Typical spark delay valves

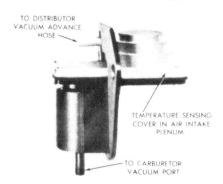

Chrysler's OSAC valve
(© Chrysler Corp)

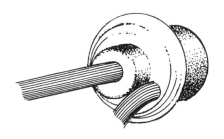

Oldsmobile vacuum reducing valve

Chrysler Orifice Spark Advance Control (OSAC)

The OSAC valve is basically the same as the temperature operated spark delay valve described above.

The OSAC valve differs in appearance and mounting location (on the firewall) from the other types of spark delay valves.

The temperature function was deleted from the OSAC valve on cars made from 15 March 1973. The new valve may be identified by a white mounting gasket and a paste-on label with the new part number.

On 1974 models, the temperature function is returned to the OSAC valve, but the valve has been moved from the firewall to the air cleaner.

Oldsmobile Vacuum Reducing Valve

Some 1973-74 Oldsmobile models use a vacuum reducing valve to prevent detonation when the coolant temperature is above 226°F.

The valve is located in the line that runs between the manifold vacuum fitting and the coolant temperature operated vacuum valve. Be careful not to confuse the vacuum reducing valve with the spark delay valve, which it resembles.

TRANSMISSION CONTROLLED SPARK (TCS) SYSTEM COMPONENTS

Manual Transmission Switch

The manual transmission switch is used to determine which gear the transmission is in, by means of a plunger on its end.

Some transmission switches are normally opened, while others are normally closed, depending upon the type of TCS system used. Do not confuse the two different types of switch if replacement is necessary.

Typical manual transmission TCS switch
(© Ford Motor Co)

The switch is threaded into the transmission housing and connected to the TCS vacuum control solenoid.

NOTE: The transmission switch resembles the back-up light switch on some models; use care not to confuse them.

Typical automatic transmission TCS switch
(© Ford Motor Co)

Automatic Transmission Switch

The automatic transmission switch is similar in function to the manual transmission switch, above. However, it determines the gear range by sensing hydraulic pressure in the transmission.

The switch is threaded into the transmission housing, except on American Motors cars made after 15 March 1973. On these cars, the switch is located on the back right-hand side of the engine and is adjustable.

NOTE: The TCS transmission switch resembles the neutral safety/back-up lamp switch used with some transmissions; use care not to confuse the two.

Vacuum Advance Solenoid

NOTE: The vacuum advance solenoid is also referred to as a "distributor modulator valve" or a "solenoid vacuum valve".

The vacuum advance solenoid is used to vent the vacuum flow to the atmosphere or to allow it to pass to the distributor vacuum unit. The solenoid is connected to the transmission switch and any other components which may be used to control vacuum advance (delay relay, temperature switch, etc.).

The solenoid may be either normally energized or de-energized, depending upon the TCS system design. Be sure to use the correct type of solenoid for replacement.

The vacuum advance solenoid is lo-

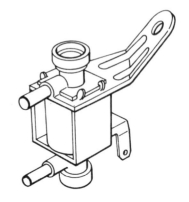

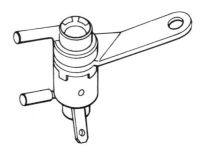

Typical vacuum advance solenoids

cated in the line which runs to the distributor vacuum unit.

CES Solenoid

The CEC solenoid is used in some

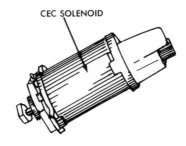

CEC SOLENOID

CEC solenoid

GM models. It has two functions: one is to regulate vacuum advance and the other is to operate as a throttle positioner during deceleration.

The CEC solenoid is mounted on the carburetor and has the distributor vacuum line running to it.

Coolant temperature switch
(© G.M. Corp)

GM Coolent Temperature Switch

The coolant temperature switch is used to override the TCS system below (or sometimes above) a specified temperature.

The switch is threaded into the coolant passages, usually at the front of the cylinder head.

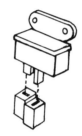

Time delay relay
(© G.M. Corp)

GM Time Delay Relay

A time delay relay is used on some GM models to provide vacuum advance for 15-20 seconds after the ignition switch is turned on, to make starting easier.

The time delay relay is wired to the CEC or vacuum advance solenoid and it is usually attached to the firewall.

Chevrolet Vacuum Delay Relay

NOTE: Do not confuse the vacuum delay relay with the time delay relay, above.

The vacuum delay relay is used on some 1972 Chevrolet models to provide a delay in vacuum advance after the transmission is shifted into high gear.

The vacuum delay relay is located inside the passenger compartment, underneath the instrument panel.

Vega TCS Relay

Vega models made in 1971-72 have an additional relay in the trans-

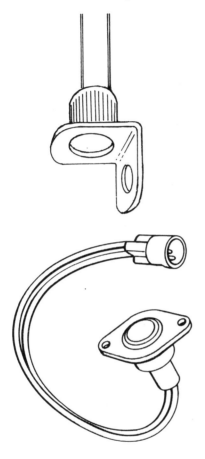

Ambient temperature switches

Ambient Temperature Switch

On some cars, an ambient temperature switch is used to override the TCS system when the outside air is below a specified temperature.

The switch is wired into the circuit which feeds power to the vacuum advance solenoid.

The ambient temperature switch is located on either front door pillar (Ford), the front crossmember (AMC), or the firewall (Chrysler).

NOTE: Because of tighter federal requirements, this switch was dropped from the ICS systems on all cars made on or after 15 March 1973.

mission controlled spark (TCS) circuit.

The relay is located on the right-hand side of the passenger compartment, toward the rear of the inner fender.

SPEED CONTROLLED SPARK (SCS) SYSTEM COMPONENTS

Speed Sensor

There are two different types of speed sensors in use. The speed sensors used by Ford and Chrysler send magnetic impulses to a control box (amplifier). The sensors used by GM and AMC use a centrifugal switch which opens, energizing the SCS circuit at a predetermined speed.

Both types are located in the speedometer cable.

Impulse-type speed sensor
(© Ford Motor Co)

Centrifugal switch-type speed sensor

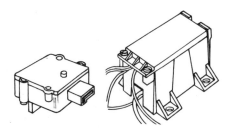

Typical-control boxes

Control Box (Amplifier)

The control box (amplifier) is used with the impulse-type speed sensor. It translates impulses from the speed sensor and sends a signal which operates the vacuum advance solenoid at a predetermined speed.

The control box is usually located on the firewall or under the instrument panel.

NOTE: On some Ford models, the amplifier and the vacuum advance solenoid are contained in one unit.

Other SCS System Components

The speed controlled spark (SCS) uses such components as vacuum advance solenoids, ambient temperature sensors, etc., which they share in common with transmission controlled spark systems (TCS).

For a description of these components, see the section above which covers TCS system components.

CARBURETOR CONTROLS

Anti-Dieseling Solenoid

NOTE: Anti-dieseling solenoids are also referred to as "throttle stop" or "idle stop" solenoids.

An anti-dieseling solenoid is used on some engines because of the high idle speed required to obtain low emission levels at idle and during deceleration. The solenoid is used to prevent dieseling (runon) after the engine is shut off.

Typical anti-dieseling solenoids

The solenoid is mounted on a bracket on the carburetor or on the intake manifold so that its plunger can come into contact with the throttle lever.

CEC Solenoid

See the description of the CEC solenoid in the section which deals with transmission controlled spark systems, above.

Pinto fuel deceleration valve
(© Ford Motor Co)

Pinto Fuel Deceleration Valve

The fuel deceleration valve used on Pintos supplies additional fuel/air mixture through the intake manifold during deceleration. The extra mixture slows the rate of deceleration, thus reducing the high level of emissions which occurs during this period.

Carburetor dashpot

Carburetor Dashpot

Some carburetors are equipped with a dashpot to keep the throttle from closing too rapidly. In addition to preventing stalling, this also aids in the reduction of emissions caused by rapid engine deceleration.

The dashpot is located on the carburetor.

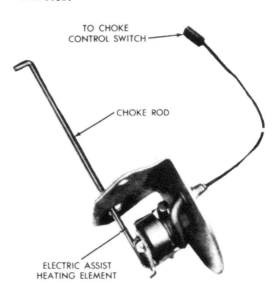

Chrysler electrically assisted choke
(© Chrysler Corp)

Ford and AMC electrically assisted choke

Electrically Assisted Choke

To shorten the duration of automatic choke operation in fairly warm weather and thus reduce emissions, a heating element is used to warm the choke.

On Ford and AMC cars, so equipped, the heater is in the choke cap. On Chrysler products, the heater is located in the choke well.

EVAPORATIVE EMISSION CONTROL (EEC) SYSTEM COMPONENTS

Charcoal Canister

The charcoal canister uses activated charcoal to store fuel vapors at idle, low speed operation, or when the engine is not running.

At high engine speeds or during acceleration, the fuel vapors are drawn out of the canister and into the air cleaner or the PCV line.

The canister is located in the engine compartment, usually in front of one of the wheel arches.

Charcoal canister—details may differ
(© Chrysler Corp)

"Pressure/Vacuum" type fuel cap
(© Chrysler Corp)

Fuel Cap

Most cars equipped with an evaporative emission control system use a "pressure/vacuum" type fuel cap.

When replacing this type of cap, it is important that the same kind of cap be used. If a non-vented fuel cap is used, it could prevent fuel delivery or cause the fuel tank to collapse; if a regular vented cap is used, it will reduce the effectiveness of the system.

Other EEC System Components

Various standpipes, separators, expansion tanks, and breather valves are used for evaporative emission control. These components are not interchangeable from one manufacturer to another; therefore, factory replacement parts should be used.

EXHAUST GAS RECIRCULATION (EGR) SYSTEM COMPONENTS

EGR Valve

A vacuum-operated exhaust gas recirculation (EGR) valve is used on many engines to route exhaust gases into the intake manifold. The addi-

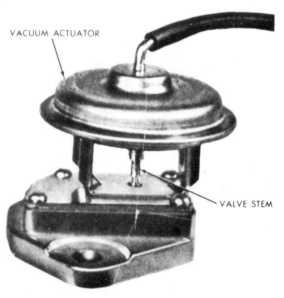

EGR valve
(© Chrysler Corp)

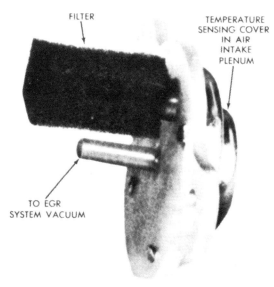

FILTER

TEMPERATURE
SENSING COVER
IN AIR
INTAKE
PLENUM

TO EGR
SYSTEM VACUUM

**Ambient temperature operated vacuum
by-pass valve**
(© Chrysler Corp)

tion of exhaust gases to the fuel/air mixture reduces peak flame temperature which helps to reduce emissions of NO_x.

The EGR valve is either mounted on the intake manifold or on a special carburetor spacer.

Temperature Overrides

Various types of temperature override switches are used to block vacuum to the EGR valve so that it will not operate under certain conditions.

Some engines use coolant temperature operated vacuum valves to regulate vacuum to the EGR valve. (See "Distributor Controls", above.) Other engines have high and/or low ambient temperature operated vacuum by-pass valves; while still others use all three to control EGR valve vacuum.

NOTE: Ambient temperature by-pass valves usually resemble those valves which perform similar functions for TCS/SCS systems. (See above.) The ambient temperature valves used on cars made on or after 15 March 1973, were either removed or covered with shrouds to make them more dependent upon engine temperature.

The coolant temperature operated vacuum valves are threaded into the water jacket of the engine, or some Chrysler products the radiator, while the ambient temperature operated valves are located in the EGR valve vacuum lines.

Chrysler Vacuum Amplifier

Some Chrysler Corp. cars use a vacuum amplifier to make up for the relatively weak EGR valve vacuum signal which is obtained at the carburetor venturi port.

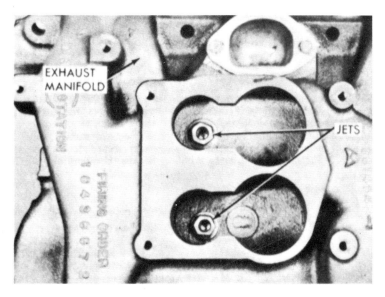

EXHAUST
MANIFOLD

JETS

Floor jets
(© Chrysler Corp)

Vacuum amplifier

Double diaphragm EGR valve used on some 1974 Buicks

Backpressure sensing device used on some 1974 AMC cars sold in California— Oldsmobile BPV is similar

The vacuum amplifier is mounted on one of the valve covers.

Floor Jets

Some Chrysler Corp. engines have jets located in the floor area, below the carburetor. Each jet has an orifice that allows a controlled amount of exhaust gas to be drawn through the jet to dilute the fuel/air mixture. The dilution of the mixture results in lower peak flame temperatures which aids in the reductions of emissions of NO_x.

Buick Double Diaphragm EGR Valve

Some 1974 Buicks have a dual diaphragm EGR valve. One hose runs to manifold vacuum, which acts to close the valve. Ported vacuum acts to open the valve in the normal manner. When manifold vacuum is high, as at idle or during deceleration, the valve stays closed. When manifold vacuum falls, ported vacuum takes over and opens the valve.

This arrangement permits fast idle without stumbling, caused by premature operation of the valve.

Exhaust Backpressure Sensor

American Motors cars use a backpressure sensing device, and Oldsmobiles use a backpressure transducer valve (BPV) on some 1974 cars sold in California.

The purpose of these devices is to prevent the EGR valve from opening at idle. Cars sold in California have higher amounts of exhaust gas recirculated, which would cause the engine to stall if the valve opened during idle.

Chapter Four

Diagnostic Roadmaps

AMERICAN MOTORS

TROUBLESHOOTING

Positive Crankcase Ventilation System

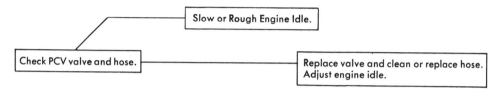

```
                    ┌─────────────────────────────┐
                    │ Slow or Rough Engine Idle.  │
                    └─────────────────────────────┘
┌──────────────────────────┐         ┌────────────────────────────────────┐
│ Check PCV valve and hose.│         │ Replace valve and clean or replace │
└──────────────────────────┘         │ hose. Adjust engine idle.          │
                                      └────────────────────────────────────┘
```

Air Guard System

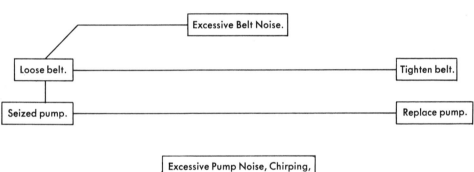

```
                    ┌─────────────────────────┐
                    │ Excessive Belt Noise.   │
                    └─────────────────────────┘
┌─────────────┐                                      ┌───────────────┐
│ Loose belt. │                                      │ Tighten belt. │
└─────────────┘                                      └───────────────┘
┌─────────────┐                                      ┌───────────────┐
│ Seized pump.│                                      │ Replace pump. │
└─────────────┘                                      └───────────────┘
```

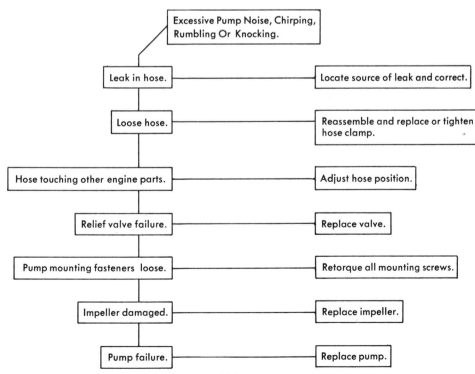

```
                    ┌───────────────────────────────────┐
                    │ Excessive Pump Noise, Chirping,    │
                    │ Rumbling Or Knocking.              │
                    └───────────────────────────────────┘
┌──────────────────┐                   ┌────────────────────────────────────┐
│ Leak in hose.    │                   │ Locate source of leak and correct. │
└──────────────────┘                   └────────────────────────────────────┘
┌──────────────────┐                   ┌────────────────────────────────────┐
│ Loose hose.      │                   │ Reassemble and replace or tighten  │
└──────────────────┘                   │ hose clamp.                        │
                                       └────────────────────────────────────┘
┌─────────────────────────────────┐    ┌────────────────────────────────────┐
│ Hose touching other engine parts.│   │ Adjust hose position.              │
└─────────────────────────────────┘    └────────────────────────────────────┘
┌──────────────────────┐               ┌────────────────────┐
│ Relief valve failure.│               │ Replace valve.     │
└──────────────────────┘               └────────────────────┘
┌──────────────────────────────┐       ┌──────────────────────────────┐
│ Pump mounting fasteners loose.│      │ Retorque all mounting screws.│
└──────────────────────────────┘       └──────────────────────────────┘
┌──────────────────────┐               ┌────────────────────┐
│ Impeller damaged.    │               │ Replace impeller.  │
└──────────────────────┘               └────────────────────┘
┌──────────────────────┐               ┌────────────────────┐
│ Pump failure.        │               │ Replace pump.      │
└──────────────────────┘               └────────────────────┘
```

AMERICAN MOTORS

Air Guard System—Cont'd

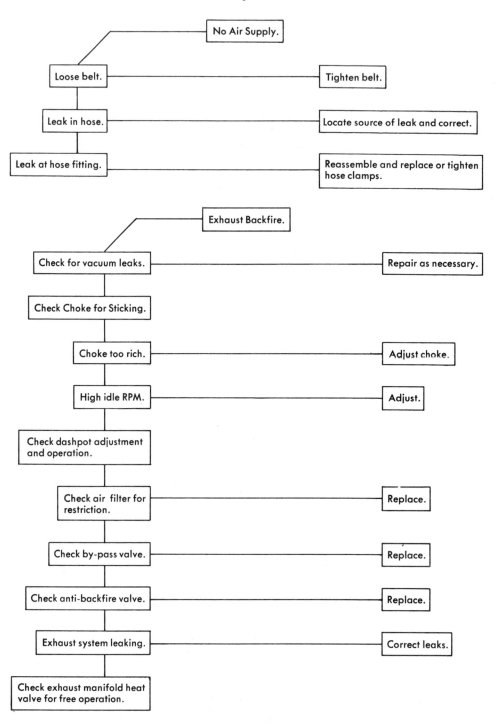

No Air Supply.

Loose belt. ———————————————— Tighten belt.

Leak in hose. ———————————————— Locate source of leak and correct.

Leak at hose fitting. ———————————————— Reassemble and replace or tighten hose clamps.

Exhaust Backfire.

Check for vacuum leaks. ———————————————— Repair as necessary.

Check Choke for Sticking.

Choke too rich. ———————————————— Adjust choke.

High idle RPM. ———————————————— Adjust.

Check dashpot adjustment and operation.

Check air filter for restriction. ———————————————— Replace.

Check by-pass valve. ———————————————— Replace.

Check anti-backfire valve. ———————————————— Replace.

Exhaust system leaking. ———————————————— Correct leaks.

Check exhaust manifold heat valve for free operation.

AMERICAN MOTORS

Air Guard System—Cont'd

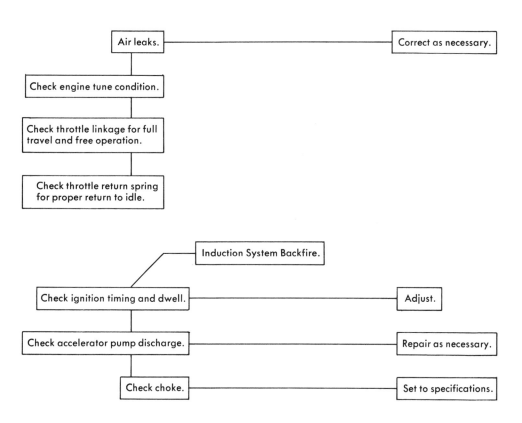

Thermostatically Controlled Air Cleaner

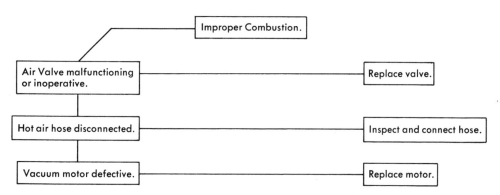

AMERICAN MOTORS

Transmission Controlled Spark System

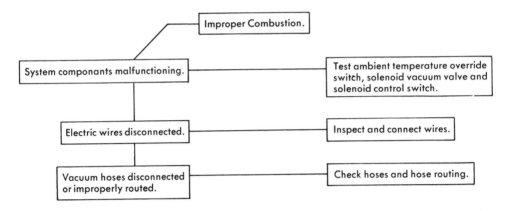

Exhaust Gas Recirculation System

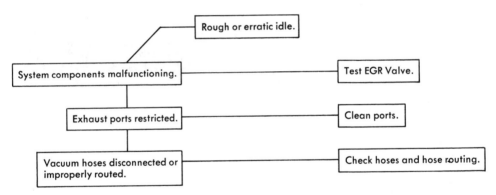

CHRYSLER

TROUBLESHOOTING

Positive Crankcase Ventilation System

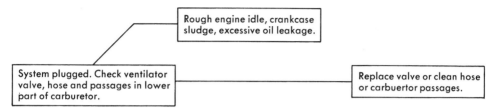

	Rough engine idle, crankcase sludge, excessive oil leakage.
System plugged. Check ventilator valve, hose and passages in lower part of carburetor.	Replace valve or clean hose or carburetor passages.

Air Injector Reactor System

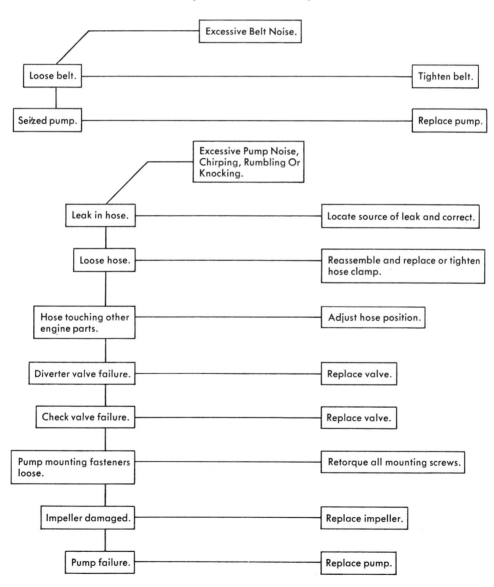

	Excessive Belt Noise.	
Loose belt.		Tighten belt.
Seized pump.		Replace pump.
	Excessive Pump Noise, Chirping, Rumbling Or Knocking.	
Leak in hose.		Locate source of leak and correct.
Loose hose.		Reassemble and replace or tighten hose clamp.
Hose touching other engine parts.		Adjust hose position.
Diverter valve failure.		Replace valve.
Check valve failure.		Replace valve.
Pump mounting fasteners loose.		Retorque all mounting screws.
Impeller damaged.		Replace impeller.
Pump failure.		Replace pump.

CHRYSLER

Air Injection Reactor System—Cont'd

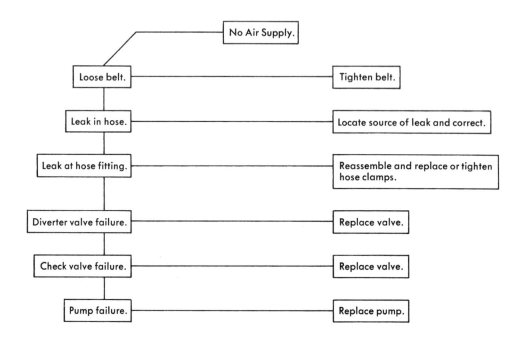

Heated Air System

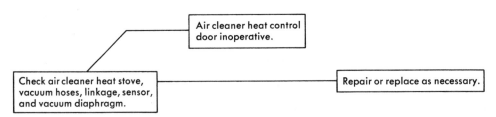

NOx System

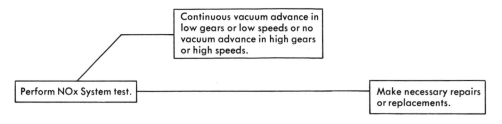

CHRYSLER

Exhaust Gas Recirculation System

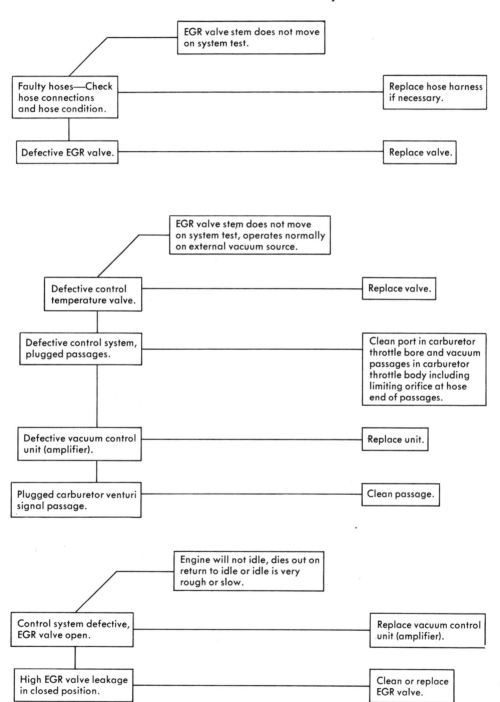

EGR valve stem does not move on system test.

Faulty hoses—Check hose connections and hose condition. → Replace hose harness if necessary.

Defective EGR valve. → Replace valve.

EGR valve stem does not move on system test, operates normally on external vacuum source.

Defective control temperature valve. → Replace valve.

Defective control system, plugged passages. → Clean port in carburetor throttle bore and vacuum passages in carburetor throttle body including limiting orifice at hose end of passages.

Defective vacuum control unit (amplifier). → Replace unit.

Plugged carburetor venturi signal passage. → Clean passage.

Engine will not idle, dies out on return to idle or idle is very rough or slow.

Control system defective, EGR valve open. → Replace vacuum control unit (amplifier).

High EGR valve leakage in closed position. → Clean or replace EGR valve.

FORD

TROUBLESHOOTING

Positive Crankcase Ventilation System

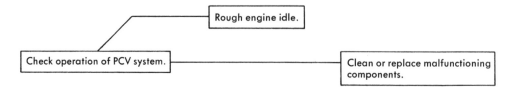

Thermactor Exhaust Emission Control System

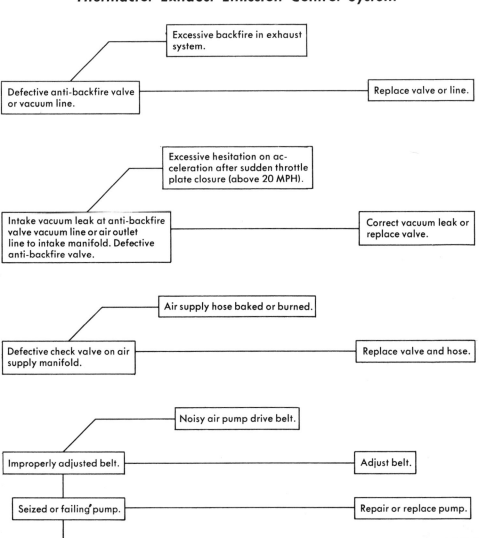

FORD

Thermactor Exhaust Emission Control System—Cont'd

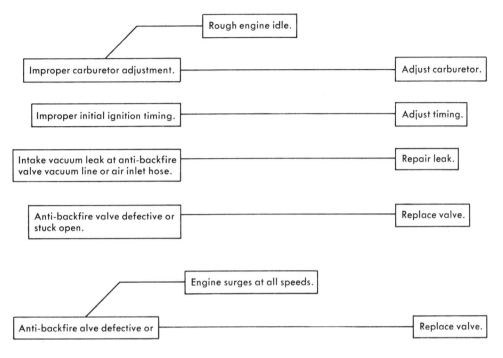

Rough engine idle.

Improper carburetor adjustment. ——— Adjust carburetor.

Improper initial ignition timing. ——— Adjust timing.

Intake vacuum leak at anti-backfire valve vacuum line or air inlet hose. ——— Repair leak.

Anti-backfire valve defective or stuck open. ——— Replace valve.

Engine surges at all speeds.

Anti-backfire alve defective or ——— Replace valve.

Air Cleaner Duct System

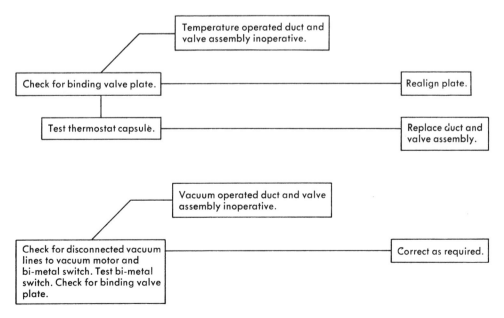

Temperature operated duct and valve assembly inoperative.

Check for binding valve plate. ——— Realign plate.

Test thermostat capsule. ——— Replace duct and valve assembly.

Vacuum operated duct and valve assembly inoperative.

Check for disconnected vacuum lines to vacuum motor and bi-metal switch. Test bi-metal switch. Check for binding valve plate. ——— Correct as required.

FORD

E.G.R. System

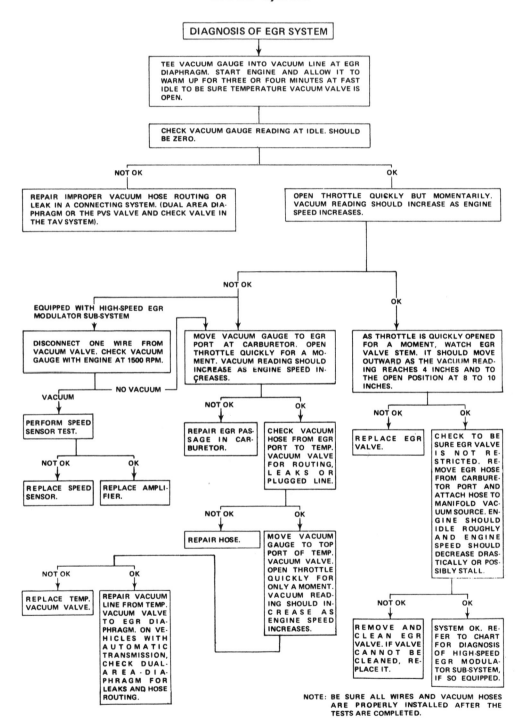

DIAGNOSIS OF EGR SYSTEM

TEE VACUUM GAUGE INTO VACUUM LINE AT EGR DIAPHRAGM. START ENGINE AND ALLOW IT TO WARM UP FOR THREE OR FOUR MINUTES AT FAST IDLE TO BE SURE TEMPERATURE VACUUM VALVE IS OPEN.

CHECK VACUUM GAUGE READING AT IDLE. SHOULD BE ZERO.

NOT OK

REPAIR IMPROPER VACUUM HOSE ROUTING OR LEAK IN A CONNECTING SYSTEM. (DUAL AREA DIA-PHRAGM OR THE PVS VALVE AND CHECK VALVE IN THE TAV SYSTEM).

OK

OPEN THROTTLE QUICKLY BUT MOMENTARILY. VACUUM READING SHOULD INCREASE AS ENGINE SPEED INCREASES.

NOT OK

OK

EQUIPPED WITH HIGH-SPEED EGR MODULATOR SUB-SYSTEM

DISCONNECT ONE WIRE FROM VACUUM VALVE. CHECK VACUUM GAUGE WITH ENGINE AT 1500 RPM.

MOVE VACUUM GAUGE TO EGR PORT AT CARBURETOR. OPEN THROTTLE QUICKLY FOR A MOMENT. VACUUM READING SHOULD INCREASE AS ENGINE SPEED IN-CREASES.

AS THROTTLE IS QUICKLY OPENED FOR A MOMENT, WATCH EGR VALVE STEM. IT SHOULD MOVE OUTWARD AS THE VACUUM READ-ING REACHES 4 INCHES AND TO THE OPEN POSITION AT 8 TO 10 INCHES.

VACUUM

NO VACUUM

PERFORM SPEED SENSOR TEST.

NOT OK

OK

REPAIR EGR PAS-SAGE IN CAR-BURETOR.

CHECK VACUUM HOSE FROM EGR PORT TO TEMP. VACUUM VALVE FOR ROUTING, LEAKS OR PLUGGED LINE.

NOT OK

OK

REPLACE EGR VALVE.

CHECK TO BE SURE EGR VALVE IS NOT RE-STRICTED. RE-MOVE EGR HOSE FROM CARBURE-TOR PORT AND ATTACH HOSE TO MANIFOLD VAC-UUM SOURCE. EN-GINE SHOULD IDLE ROUGHLY AND ENGINE SPEED SHOULD DECREASE DRAS-TICALLY OR POS-SIBLY STALL.

NOT OK

OK

REPLACE SPEED SENSOR.

REPLACE AMPLI-FIER.

NOT OK

OK

REPAIR HOSE.

MOVE VACUUM GAUGE TO TOP PORT OF TEMP. VACUUM VALVE. OPEN THROTTLE QUICKLY FOR ONLY A MOMENT. VACUUM READ-ING SHOULD IN-CREASE AS ENGINE SPEED INCREASES.

NOT OK

OK

REPLACE TEMP. VACUUM VALVE.

REPAIR VACUUM LINE FROM TEMP. VACUUM VALVE TO EGR DIA-PHRAGM. ON VE-HICLES WITH AUTOMATIC TRANSMISSION, CHECK DUAL-AREA-DIA-PHRAGM FOR LEAKS AND HOSE ROUTING.

NOT OK

OK

REMOVE AND CLEAN EGR VALVE. IF VALVE CANNOT BE CLEANED, RE-PLACE IT.

SYSTEM OK. RE-FER TO CHART FOR DIAGNOSIS OF HIGH-SPEED EGR MODULA-TOR SUB-SYSTEM, IF SO EQUIPPED.

NOTE: BE SURE ALL WIRES AND VACUUM HOSES ARE PROPERLY INSTALLED AFTER THE TESTS ARE COMPLETED.

FORD

High-Speed EGR Modulator Sub-System

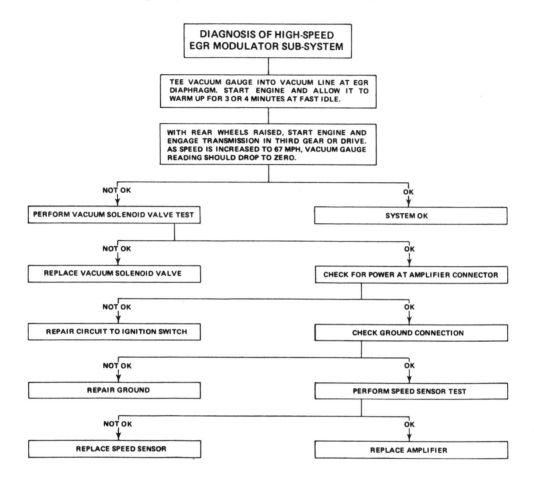

FORD

TRS System

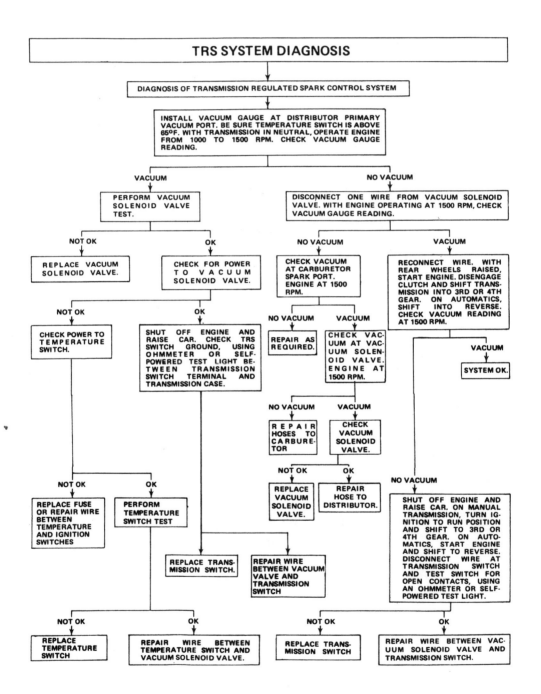

TRS SYSTEM DIAGNOSIS

DIAGNOSIS OF TRANSMISSION REGULATED SPARK CONTROL SYSTEM

INSTALL VACUUM GAUGE AT DISTRIBUTOR PRIMARY VACUUM PORT. BE SURE TEMPERATURE SWITCH IS ABOVE 65ºF. WITH TRANSMISSION IN NEUTRAL, OPERATE ENGINE FROM 1000 TO 1500 RPM. CHECK VACUUM GAUGE READING.

VACUUM

PERFORM VACUUM SOLENOID VALVE TEST.

NOT OK

REPLACE VACUUM SOLENOID VALVE.

OK

CHECK FOR POWER TO VACUUM SOLENOID VALVE.

NOT OK

CHECK POWER TO TEMPERATURE SWITCH.

OK

SHUT OFF ENGINE AND RAISE CAR. CHECK TRS SWITCH GROUND, USING OHMMETER OR SELF-POWERED TEST LIGHT BETWEEN TRANSMISSION SWITCH TERMINAL AND TRANSMISSION CASE.

NOT OK

REPLACE FUSE OR REPAIR WIRE BETWEEN TEMPERATURE AND IGNITION SWITCHES

OK

PERFORM TEMPERATURE SWITCH TEST

NOT OK

REPLACE TEMPERATURE SWITCH

OK

REPAIR WIRE BETWEEN TEMPERATURE SWITCH AND VACUUM SOLENOID VALVE.

REPLACE TRANSMISSION SWITCH.

NO VACUUM

DISCONNECT ONE WIRE FROM VACUUM SOLENOID VALVE. WITH ENGINE OPERATING AT 1500 RPM, CHECK VACUUM GAUGE READING.

NO VACUUM

CHECK VACUUM AT CARBURETOR SPARK PORT. ENGINE AT 1500 RPM.

NO VACUUM

REPAIR AS REQUIRED.

VACUUM

CHECK VACUUM AT VACUUM SOLENOID VALVE. ENGINE AT 1500 RPM.

NO VACUUM

REPAIR HOSES TO CARBURETOR

NOT OK

REPLACE VACUUM SOLENOID VALVE.

VACUUM

CHECK VACUUM SOLENOID VALVE.

OK

REPAIR HOSE TO DISTRIBUTOR.

REPAIR WIRE BETWEEN VACUUM VALVE AND TRANSMISSION SWITCH

VACUUM

RECONNECT WIRE. WITH REAR WHEELS RAISED, START ENGINE. DISENGAGE CLUTCH AND SHIFT TRANSMISSION INTO 3RD OR 4TH GEAR. ON AUTOMATICS, SHIFT INTO REVERSE. CHECK VACUUM READING AT 1500 RPM.

VACUUM

SYSTEM OK.

NO VACUUM

SHUT OFF ENGINE AND RAISE CAR. ON MANUAL TRANSMISSION, TURN IGNITION TO RUN POSITION AND SHIFT TO 3RD OR 4TH GEAR. ON AUTOMATICS, START ENGINE AND SHIFT TO REVERSE. DISCONNECT WIRE AT TRANSMISSION SWITCH AND TEST SWITCH FOR OPEN CONTACTS, USING AN OHMMETER OR SELF-POWERED TEST LIGHT.

NOT OK

REPLACE TRANSMISSION SWITCH

OK

REPAIR WIRE BETWEEN VACUUM SOLENOID VALVE AND TRANSMISSION SWITCH.

FORD

TRS+1 System

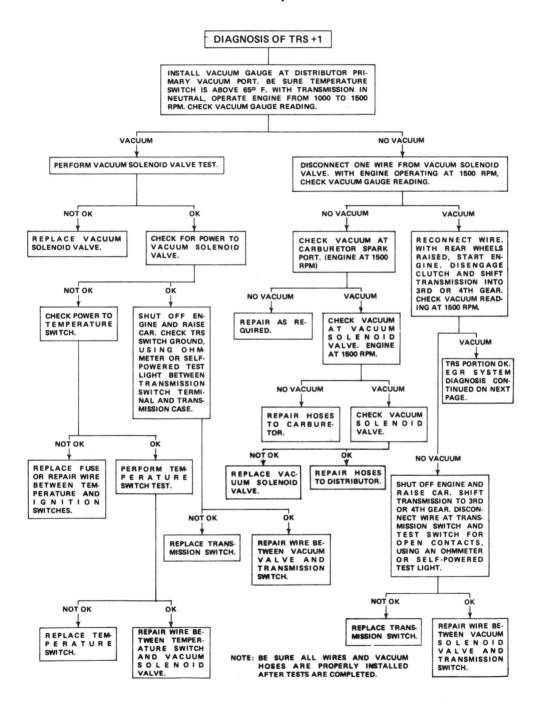

DIAGNOSIS OF TRS +1

INSTALL VACUUM GAUGE AT DISTRIBUTOR PRIMARY VACUUM PORT. BE SURE TEMPERATURE SWITCH IS ABOVE 65° F. WITH TRANSMISSION IN NEUTRAL, OPERATE ENGINE FROM 1000 TO 1500 RPM. CHECK VACUUM GAUGE READING.

VACUUM

PERFORM VACUUM SOLENOID VALVE TEST.

NOT OK

REPLACE VACUUM SOLENOID VALVE.

OK

CHECK FOR POWER TO VACUUM SOLENOID VALVE.

NOT OK

CHECK POWER TO TEMPERATURE SWITCH.

OK

SHUT OFF ENGINE AND RAISE CAR. CHECK TRS SWITCH GROUND, USING OHMMETER OR SELF-POWERED TEST LIGHT BETWEEN TRANSMISSION SWITCH TERMINAL AND TRANSMISSION CASE.

NOT OK

REPLACE FUSE OR REPAIR WIRE BETWEEN TEMPERATURE AND IGNITION SWITCHES.

OK

PERFORM TEMPERATURE SWITCH TEST.

NOT OK

REPLACE TEMPERATURE SWITCH.

OK

REPAIR WIRE BETWEEN TEMPERATURE SWITCH AND VACUUM SOLENOID VALVE.

NO VACUUM

DISCONNECT ONE WIRE FROM VACUUM SOLENOID VALVE. WITH ENGINE OPERATING AT 1500 RPM, CHECK VACUUM GAUGE READING.

NO VACUUM

CHECK VACUUM AT CARBURETOR SPARK PORT. (ENGINE AT 1500 RPM)

NO VACUUM

REPAIR AS REQUIRED.

VACUUM

CHECK VACUUM AT VACUUM SOLENOID VALVE. ENGINE AT 1500 RPM.

NO VACUUM

REPAIR HOSES TO CARBURETOR.

VACUUM

CHECK VACUUM SOLENOID VALVE.

NOT OK

REPLACE VACUUM SOLENOID VALVE.

OK

REPAIR HOSES TO DISTRIBUTOR.

NOT OK

REPLACE TRANSMISSION SWITCH.

OK

REPAIR WIRE BETWEEN VACUUM VALVE AND TRANSMISSION SWITCH.

VACUUM

RECONNECT WIRE. WITH REAR WHEELS RAISED, START ENGINE. DISENGAGE CLUTCH AND SHIFT TRANSMISSION INTO 3RD OR 4TH GEAR. CHECK VACUUM READING AT 1500 RPM.

VACUUM

TRS PORTION OK. EGR SYSTEM DIAGNOSIS CONTINUED ON NEXT PAGE.

NO VACUUM

SHUT OFF ENGINE AND RAISE CAR. SHIFT TRANSMISSION TO 3RD OR 4TH GEAR. DISCONNECT WIRE AT TRANSMISSION SWITCH AND TEST SWITCH FOR OPEN CONTACTS, USING AN OHMMETER OR SELF-POWERED TEST LIGHT.

NOT OK

REPLACE TRANSMISSION SWITCH.

OK

REPAIR WIRE BETWEEN VACUUM SOLENOID VALVE AND TRANSMISSION SWITCH.

NOTE: BE SURE ALL WIRES AND VACUUM HOSES ARE PROPERLY INSTALLED AFTER TESTS ARE COMPLETED.

FORD

TRS+1 System—Cont'd

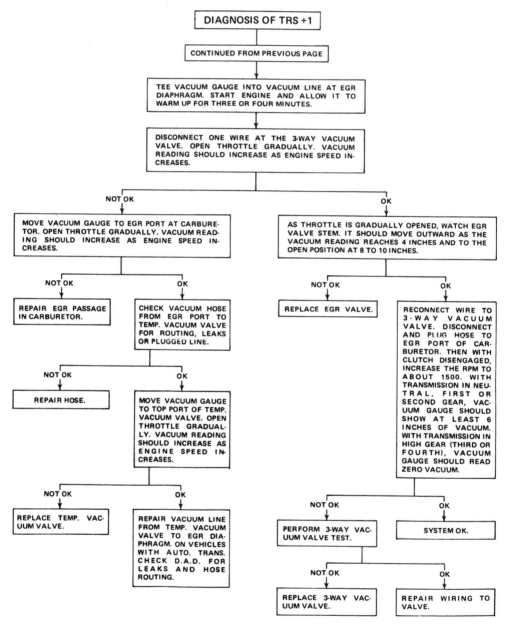

NOTE: BE SURE ALL WIRES AND VACUUM HOSES ARE
PROPERLY INSTALLED AFTER TESTS ARE COMPLETED.

FORD

ECS System

ESC SYSTEM DIAGNOSIS

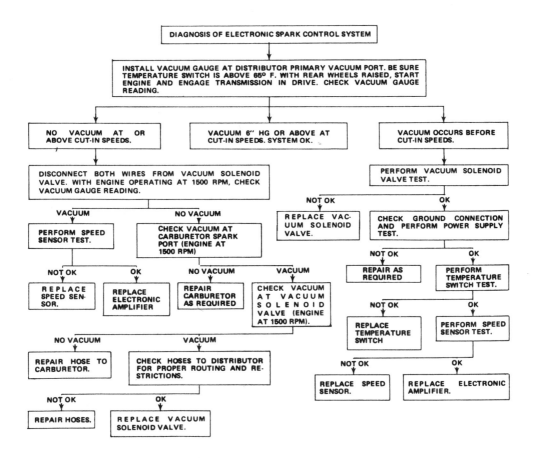

FORD

Delay Vacuum By-Pass System

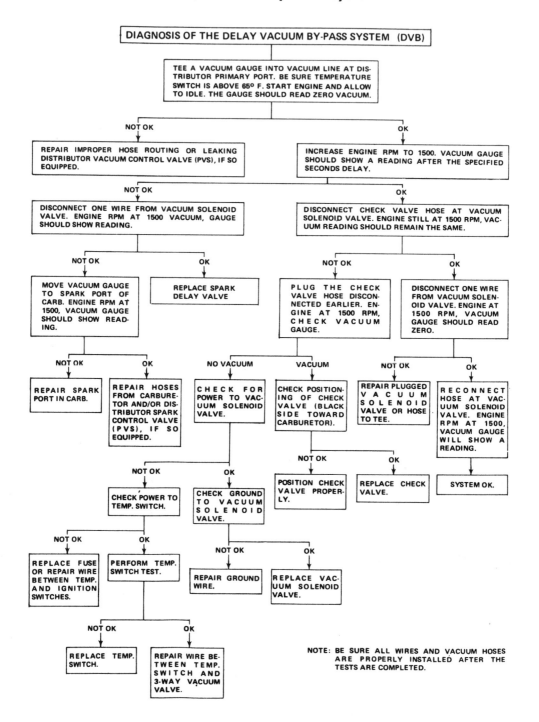

FORD

Temperature Actuated System

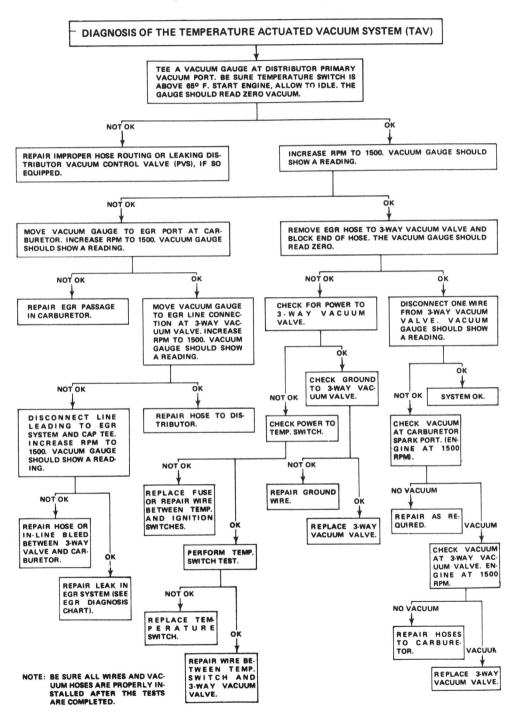

DIAGNOSIS OF THE TEMPERATURE ACTUATED VACUUM SYSTEM (TAV)

TEE A VACUUM GAUGE AT DISTRIBUTOR PRIMARY VACUUM PORT. BE SURE TEMPERATURE SWITCH IS ABOVE 65° F. START ENGINE, ALLOW TO IDLE. THE GAUGE SHOULD READ ZERO VACUUM.

NOT OK

REPAIR IMPROPER HOSE ROUTING OR LEAKING DISTRIBUTOR VACUUM CONTROL VALVE (PVS), IF SO EQUIPPED.

OK

INCREASE RPM TO 1500. VACUUM GAUGE SHOULD SHOW A READING.

NOT OK

MOVE VACUUM GAUGE TO EGR PORT AT CARBURETOR. INCREASE RPM TO 1500. VACUUM GAUGE SHOULD SHOW A READING.

OK

REMOVE EGR HOSE TO 3-WAY VACUUM VALVE AND BLOCK END OF HOSE. THE VACUUM GAUGE SHOULD READ ZERO.

NOT OK

REPAIR EGR PASSAGE IN CARBURETOR.

OK

MOVE VACUUM GAUGE TO EGR LINE CONNECTION AT 3-WAY VACUUM VALVE. INCREASE RPM TO 1500. VACUUM GAUGE SHOULD SHOW A READING.

NOT OK

CHECK FOR POWER TO 3-WAY VACUUM VALVE.

OK

CHECK GROUND TO 3-WAY VACUUM VALVE.

OK

DISCONNECT ONE WIRE FROM 3-WAY VACUUM VALVE. VACUUM GAUGE SHOULD SHOW A READING.

OK

NOT OK   SYSTEM OK.

NOT OK

DISCONNECT LINE LEADING TO EGR SYSTEM AND CAP TEE. INCREASE RPM TO 1500. VACUUM GAUGE SHOULD SHOW A READING.

OK

REPAIR HOSE TO DISTRIBUTOR.

NOT OK

CHECK POWER TO TEMP. SWITCH.

NOT OK

CHECK VACUUM AT CARBURETOR SPARK PORT. (ENGINE AT 1500 RPM).

NOT OK

REPAIR HOSE OR IN-LINE BLEED BETWEEN 3-WAY VALVE AND CARBURETOR.

OK

NOT OK

REPLACE FUSE OR REPAIR WIRE BETWEEN TEMP. AND IGNITION SWITCHES.

NOT OK

REPAIR GROUND WIRE.

OK

NO VACUUM

REPAIR AS REQUIRED.

VACUUM

REPAIR LEAK IN EGR SYSTEM (SEE EGR DIAGNOSIS CHART).

OK

PERFORM TEMP. SWITCH TEST.

OK

REPLACE 3-WAY VACUUM VALVE.

CHECK VACUUM AT 3-WAY VACUUM VALVE. ENGINE AT 1500 RPM.

NOT OK

REPLACE TEMPERATURE SWITCH.

OK

NO VACUUM

REPAIR HOSES TO CARBURETOR.

VACUUM

NOTE: BE SURE ALL WIRES AND VACUUM HOSES ARE PROPERLY INSTALLED AFTER THE TESTS ARE COMPLETED.

REPAIR WIRE BETWEEN TEMP. SWITCH AND 3-WAY VACUUM VALVE.

REPLACE 3-WAY VACUUM VALVE.

FORD

Cold Temperature Actuated System

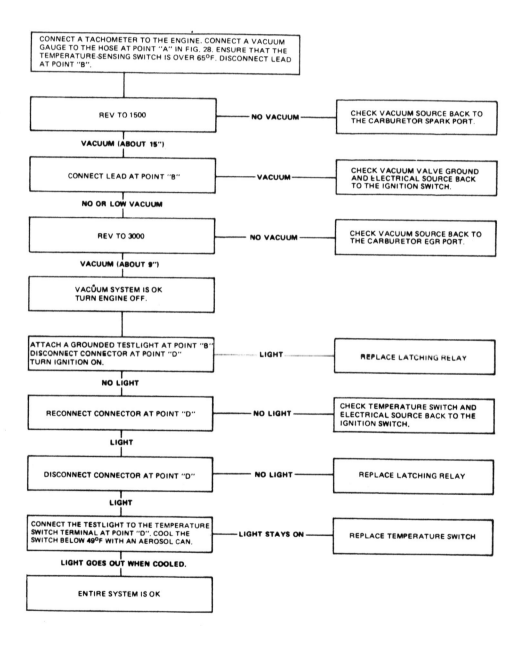

FORD

FORD PINTO
TROUBLESHOOTING

Positive Crankcase Ventilation System

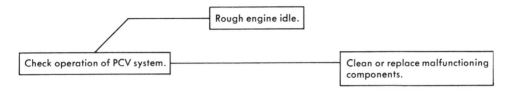

Air Cleaner Duct System

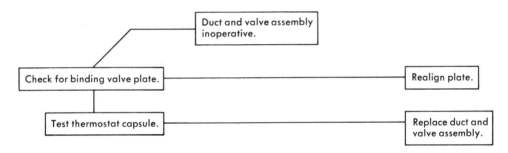

Decel Valve

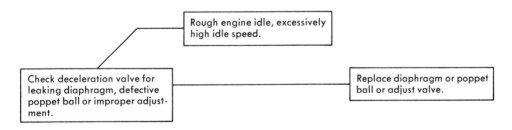

GENERAL MOTORS

TROUBLESHOOTING

Positive Crankcase Ventilation System

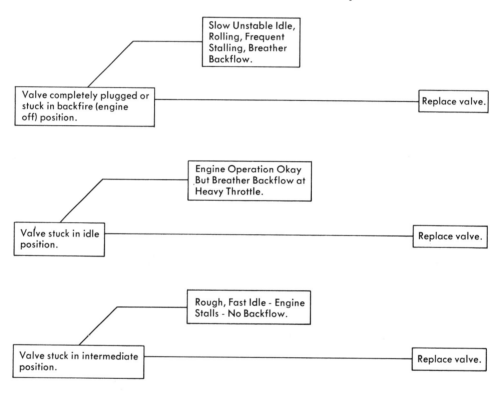

Slow Unstable Idle, Rolling, Frequent Stalling, Breather Backflow.

Valve completely plugged or stuck in backfire (engine off) position. — Replace valve.

Engine Operation Okay But Breather Backflow at Heavy Throttle.

Valve stuck in idle position. — Replace valve.

Rough, Fast Idle - Engine Stalls - No Backflow.

Valve stuck in intermediate position. — Replace valve.

Heated Air System

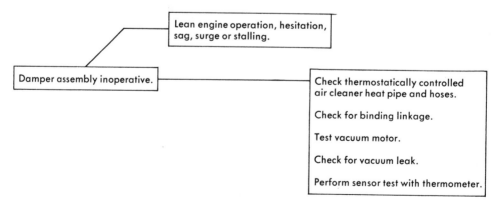

Lean engine operation, hesitation, sag, surge or stalling.

Damper assembly inoperative. —
Check thermostatically controlled air cleaner heat pipe and hoses.

Check for binding linkage.

Test vacuum motor.

Check for vacuum leak.

Perform sensor test with thermometer.

GENERAL MOTORS

Transmission Controlled Spark

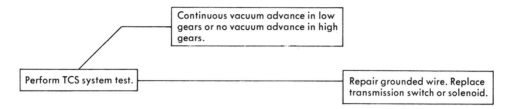

Continuous vacuum advance in low gears or no vacuum advance in high gears.

Perform TCS system test.

Repair grounded wire. Replace transmission switch or solenoid.

Air Injection Reactor System

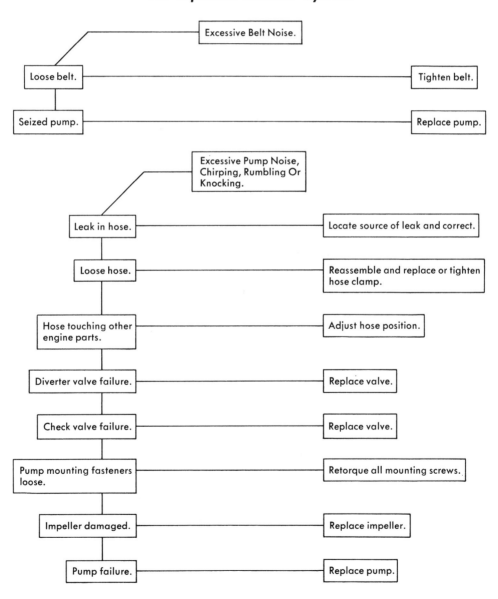

Excessive Belt Noise.

Loose belt. — Tighten belt.

Seized pump. — Replace pump.

Excessive Pump Noise, Chirping, Rumbling Or Knocking.

Leak in hose. — Locate source of leak and correct.

Loose hose. — Reassemble and replace or tighten hose clamp.

Hose touching other engine parts. — Adjust hose position.

Diverter valve failure. — Replace valve.

Check valve failure. — Replace valve.

Pump mounting fasteners loose. — Retorque all mounting screws.

Impeller damaged. — Replace impeller.

Pump failure. — Replace pump.

GENERAL MOTORS

Air Injection Reactor System—Cont'd

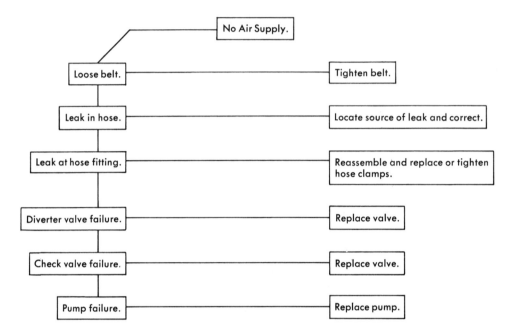

Exhaust Gas Recirculation System

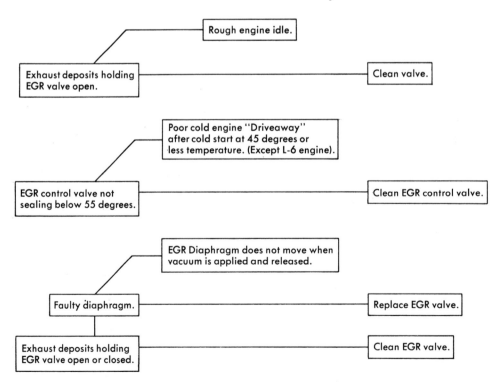

GENERAL MOTORS

Speed Controlled Switch System
Cadillac

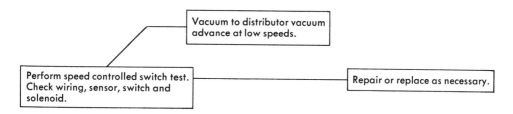

Evaporative Emission Control System

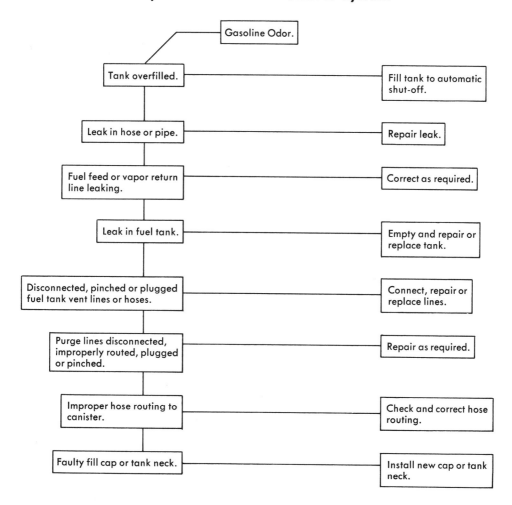

GENERAL MOTORS

Evaporative Emission Control System—Cont'd

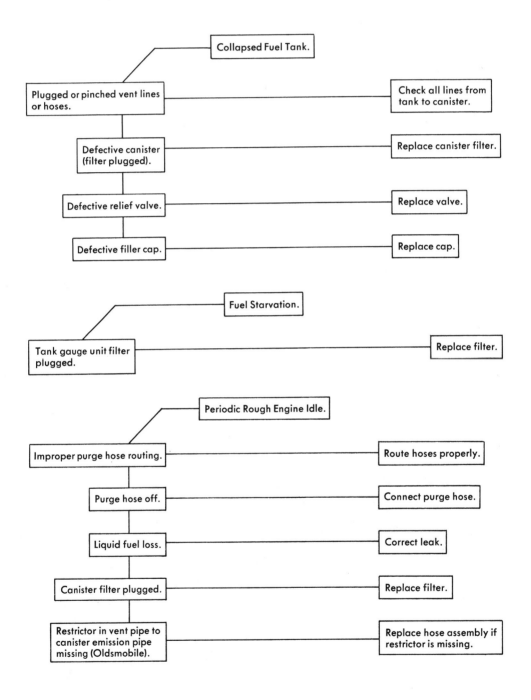

Collapsed Fuel Tank.

Plugged or pinched vent lines or hoses. — Check all lines from tank to canister.

Defective canister (filter plugged). — Replace canister filter.

Defective relief valve. — Replace valve.

Defective filler cap. — Replace cap.

Fuel Starvation.

Tank gauge unit filter plugged. — Replace filter.

Periodic Rough Engine Idle.

Improper purge hose routing. — Route hoses properly.

Purge hose off. — Connect purge hose.

Liquid fuel loss. — Correct leak.

Canister filter plugged. — Replace filter.

Restrictor in vent pipe to canister emission pipe missing (Oldsmobile). — Replace hose assembly if restrictor is missing.

GENERAL MOTORS

Vega T.C.S. 1971-72

TROUBLE SHOOTING GUIDE
VEGA ENGINE

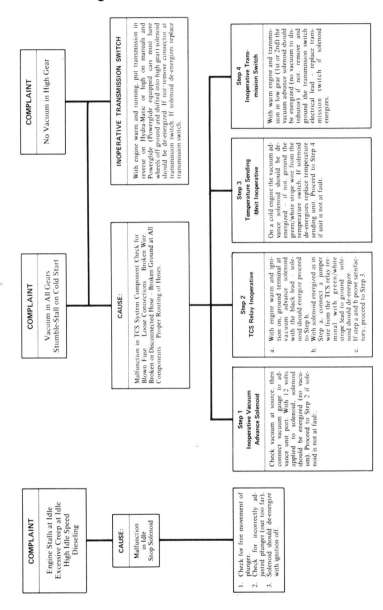

COMPLAINT

No Vacuum in High Gear

COMPLAINT

Vacuum in All Gears
Stumble-Stall on Cold Start

INOPERATIVE TRANSMISSION SWITCH

With engine warm and running, put transmission in reverse on Hydra-Matic or high on manuals and Powerglide (Powerglide equipped cars must have wheels off ground and shifted into high gear) solenoid should be de-energized. If not remove connector at transmission switch. If solenoid de-energizes replace transmission switch.

CAUSE:

Malfunction in TCS System Component Check for Blown Fuse   Loose Connections   Broken Wire Broken or Disconnected Hose – Broken Ground at All Components   Proper Routing of Hoses.

Step 4
Inoperative Trans-
mission Switch

With warm engine and transmission in low gear (1st or 2nd) the vacuum advance solenoid should be energized (no vacuum to distributor) if not, remove and ground the transmission switch electrical lead – replace transmission switch if solenoid energizes.

Step 3
Temperature Sending
Unit Inoperative

On a cold engine the vacuum advance solenoid should be de-energized – if not, ground the green/white stripe wire from the temperature switch. If solenoid de-energizes replace temperature sending unit. Proceed to Step 4 if unit is not at fault.

Step 2
TCS Relay Inoperative

a. With engine warm and ignition on, ground terminal at vacuum advance solenoid with the black lead – solenoid should energize proceed to Step b.
b. With solenoid energized as in Step a, connect a jumper wire from the TCS relay terminal with green/white stripe lead to ground solenoid should de-energize.
c. If step a and b prove satisfactory, proceed to Step 3.

Step 1
Inoperative Vacuum
Advance Solenoid

Check vacuum at source, then connect vacuum gauge to advance unit port. With 12 volts applied to solenoid, solenoid should be energized (no vacuum). Proceed to Step 2 if solenoid is not at fault.

COMPLAINT

Engine Stalls at Idle
Excessive Creep at Idle
High Idle Speed
Dieseling

CAUSE:

Malfunction
in Idle
Stop Solenoid

1. Check for free movement of plunger.
2. Check for incorrectly adjusted plunger (out too far).
3. Solenoid should de-energize with ignition off.

GENERAL MOTORS

Vega T.C.S. 1973-74

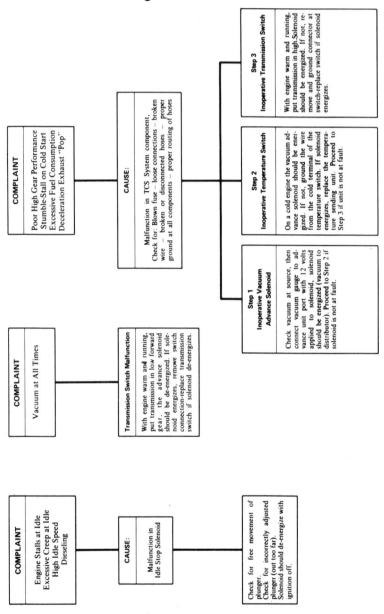

COMPLAINT

Poor High Gear Performance
Stumble-Stall on Cold Start
Excessive Fuel Consumption
Deceleration Exhaust "Pop"

CAUSE:

Malfunction in TCS System component,
Check for: Blown fuse — loose connections — broken
wire — broken or disconnected hoses — proper
ground at all components — proper routing of hoses

Step 1
Inoperative Vacuum Advance Solenoid

Check vacuum at source, then connect vacuum gauge to advance unit port with 12 volts applied to solenoid, solenoid should be energized (vacuum to distributor). Proceed to Step 2 if solenoid is not at fault.

Step 2
Inoperative Temperature Switch

On a cold engine the vacuum advance solenoid should be energized. If not, ground the wire from the cold terminal of the temperature switch. If solenoid energizes, replace the temperature sending unit. Proceed to Step 3 if unit is not at fault.

Step 3
Inoperative Transmission Switch

With engine warm and running, put transmission in high. Solenoid should be energized. If not, remove and ground connector at switch-replace switch if solenoid energizes.

COMPLAINT

Vacuum at All Times

Transmission Switch Malfunction

With engine warm and running, put transmission in low forward gear. the advance solenoid should be de-energized. If solenoid energizes, remove switch connection-replace transmission switch if solenoid de-energizes.

COMPLAINT

Engine Stalls at Idle
Excessive Creep at Idle
High Idle Speed
Dieseling

CAUSE:

Malfunction in
Idle Stop Solenoid

Check for free movement of plunger.
Check for incorrectly adjusted plunger (out too far).
Solenoid should de-energize with ignition off.

VOLKSWAGEN

TROUBLESHOOTING

Throttle Positioner

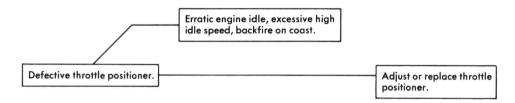

Erratic engine idle, excessive high idle speed, backfire on coast.

Defective throttle positioner.

Adjust or replace throttle positioner.

Exhaust Gas Recirculation System

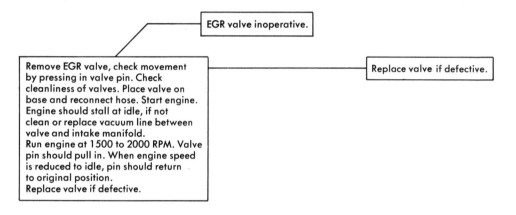

EGR valve inoperative.

Remove EGR valve, check movement by pressing in valve pin. Check cleanliness of valves. Place valve on base and reconnect hose. Start engine. Engine should stall at idle, if not clean or replace vacuum line between valve and intake manifold.
Run engine at 1500 to 2000 RPM. Valve pin should pull in. When engine speed is reduced to idle, pin should return to original position.
Replace valve if defective.

Replace valve if defective.

TOYOTA

TROUBLESHOOTING

Position Crankcase Ventilation System

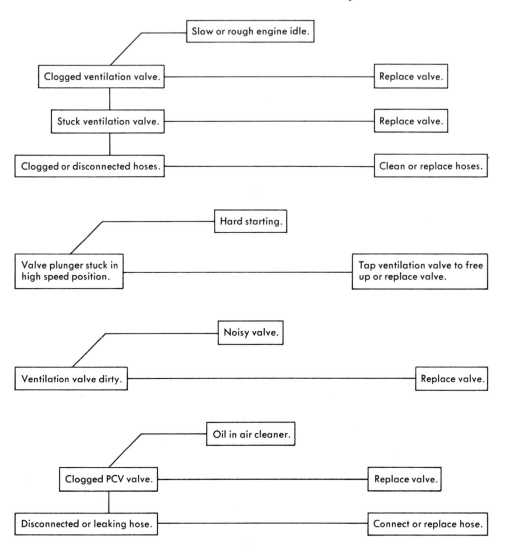

TOYOTA

Air Injection System

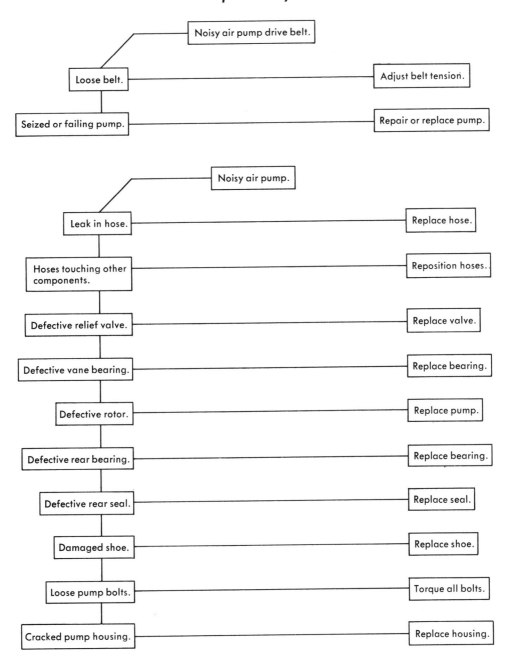

TOYOTA

Air Injection—Cont'd

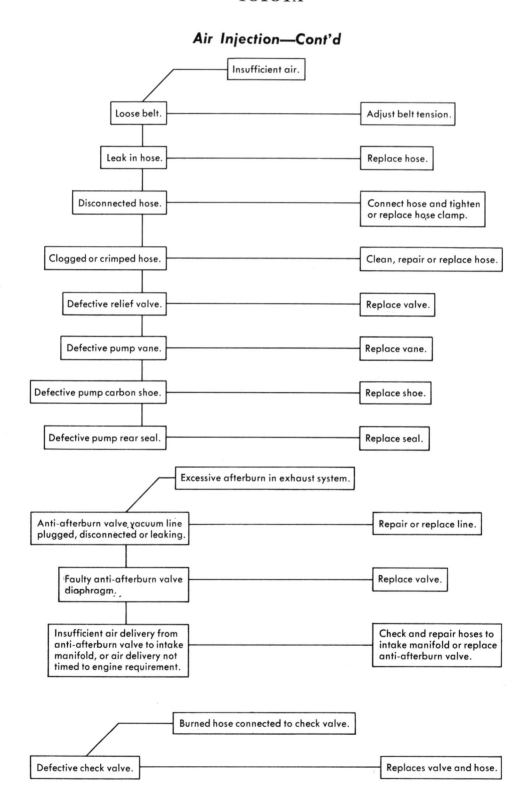

Insufficient air.

Loose belt. ——— Adjust belt tension.

Leak in hose. ——— Replace hose.

Disconnected hose. ——— Connect hose and tighten or replace hose clamp.

Clogged or crimped hose. ——— Clean, repair or replace hose.

Defective relief valve. ——— Replace valve.

Defective pump vane. ——— Replace vane.

Defective pump carbon shoe. ——— Replace shoe.

Defective pump rear seal. ——— Replace seal.

Excessive afterburn in exhaust system.

Anti-afterburn valve, vacuum line plugged, disconnected or leaking. ——— Repair or replace line.

Faulty anti-afterburn valve diaphragm. ——— Replace valve.

Insufficient air delivery from anti-afterburn valve to intake manifold, or air delivery not timed to engine requirement. ——— Check and repair hoses to intake manifold or replace anti-afterburn valve.

Burned hose connected to check valve.

Defective check valve. ——— Replaces valve and hose.

TOYOTA

Air Injection—Cont'd

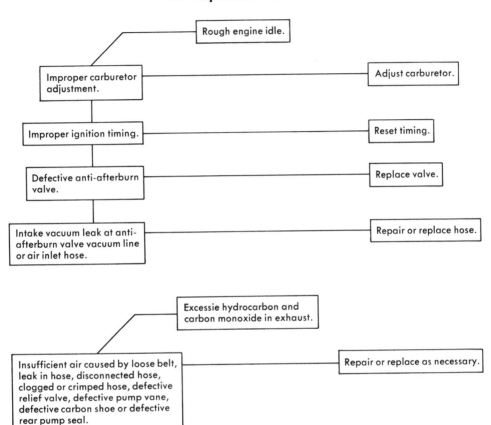

TOYOTA

Throttle Positioner System

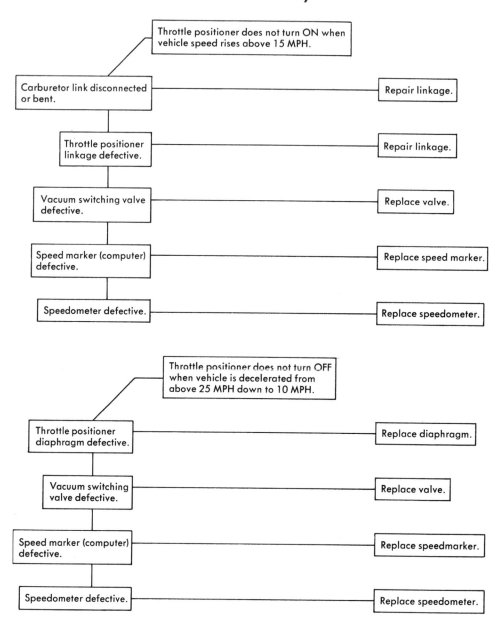

TOYOTA

Transmission Controlled Spark System

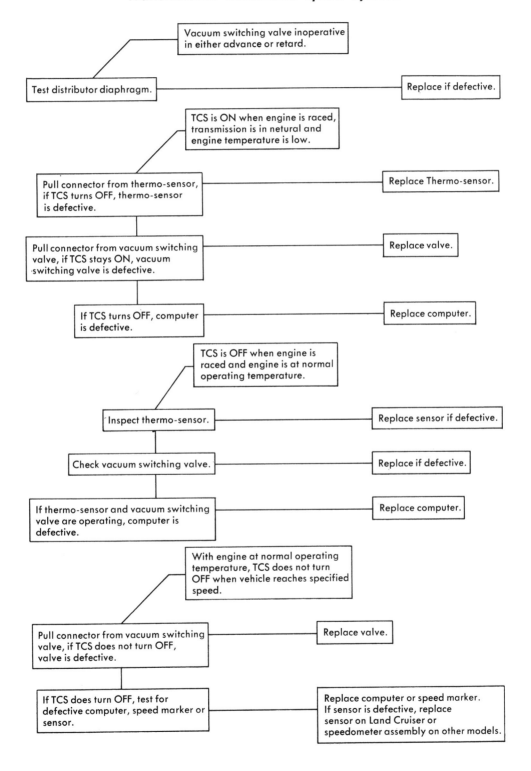

Vacuum switching valve inoperative in either advance or retard.

Test distributor diaphragm. — Replace if defective.

TCS is ON when engine is raced, transmission is in netural and engine temperature is low.

Pull connector from thermo-sensor, if TCS turns OFF, thermo-sensor is defective. — Replace Thermo-sensor.

Pull connector from vacuum switching valve, if TCS stays ON, vacuum ·switching valve is defective. — Replace valve.

If TCS turns OFF, computer is defective. — Replace computer.

TCS is OFF when engine is raced and engine is at normal operating temperature.

Inspect thermo-sensor. — Replace sensor if defective.

Check vacuum switching valve. — Replace if defective.

If thermo-sensor and vacuum switching valve are operating, computer is defective. — Replace computer.

With engine at normal operating temperature, TCS does not turn OFF when vehicle reaches specified speed.

Pull connector from vacuum switching valve, if TCS does not turn OFF, valve is defective. — Replace valve.

If TCS does turn OFF, test for defective computer, speed marker or sensor. — Replace computer or speed marker. If sensor is defective, replace sensor on Land Cruiser or speedometer assembly on other models.

TOYOTA

Mixture Control System

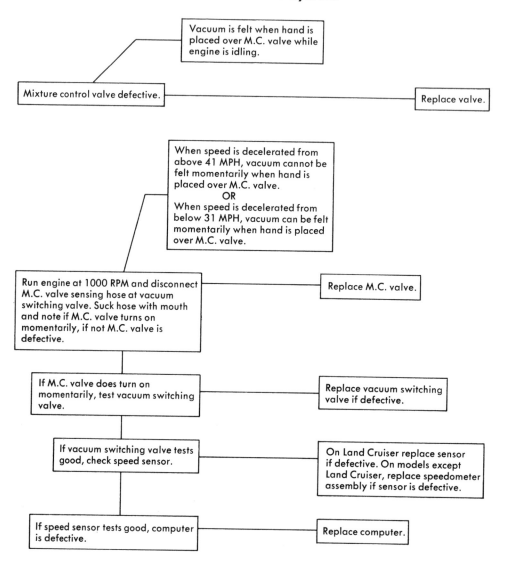

Vacuum is felt when hand is placed over M.C. valve while engine is idling.

Mixture control valve defective.

Replace valve.

When speed is decelerated from above 41 MPH, vacuum cannot be felt momentarily when hand is placed over M.C. valve.
OR
When speed is decelerated from below 31 MPH, vacuum can be felt momentarily when hand is placed over M.C. valve.

Run engine at 1000 RPM and disconnect M.C. valve sensing hose at vacuum switching valve. Suck hose with mouth and note if M.C. valve turns on momentarily, if not M.C. valve is defective.

Replace M.C. valve.

If M.C. valve does turn on momentarily, test vacuum switching valve.

Replace vacuum switching valve if defective.

If vacuum switching valve tests good, check speed sensor.

On Land Cruiser replace sensor if defective. On models except Land Cruiser, replace speedometer assembly if sensor is defective.

If speed sensor tests good, computer is defective.

Replace computer.

TOYOTA

Charcoal Canister Storage System

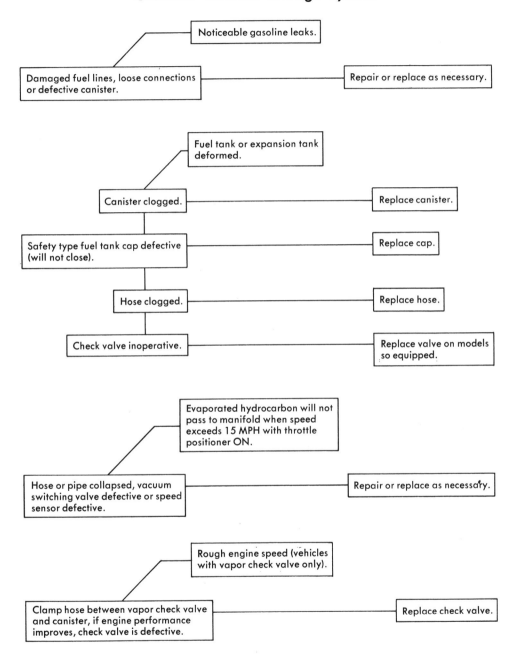

Noticeable gasoline leaks.

Damaged fuel lines, loose connections or defective canister. — Repair or replace as necessary.

Fuel tank or expansion tank deformed.

Canister clogged. — Replace canister.

Safety type fuel tank cap defective (will not close). — Replace cap.

Hose clogged. — Replace hose.

Check valve inoperative. — Replace valve on models so equipped.

Evaporated hydrocarbon will not pass to manifold when speed exceeds 15 MPH with throttle positioner ON.

Hose or pipe collapsed, vacuum switching valve defective or speed sensor defective. — Repair or replace as necessary.

Rough engine speed (vehicles with vapor check valve only).

Clamp hose between vapor check valve and canister, if engine performance improves, check valve is defective. — Replace check valve.

TOYOTA

Exhaust Gas Recirculation System

EGR System Inspection Chart (18R-C & F engines)

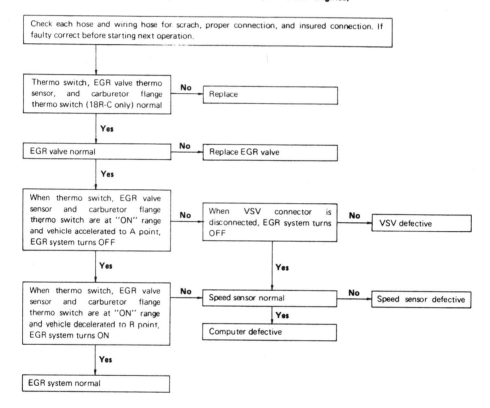

TOYOTA

Exhaust Gas Recirculation System—Cont'd

EGR System Inspection Chart (4M engine)

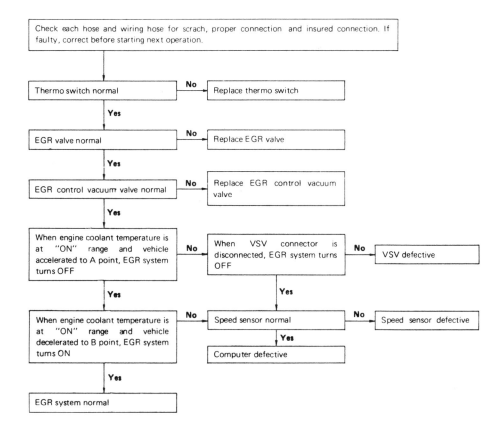

TOYOTA

Exhaust Gas Recirculation System—Cont'd

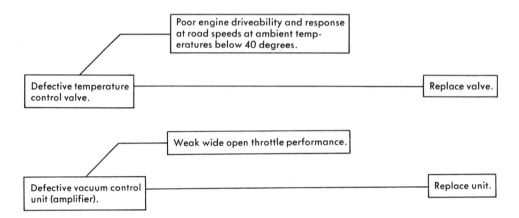

Evaporator Control System

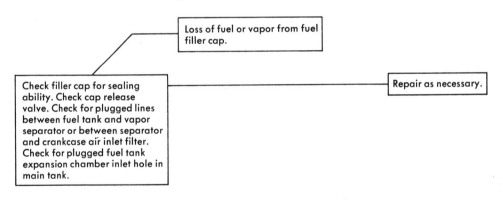

DATSUN

TROUBLESHOOTING

Positive Crankcase Ventilation System

Air Injection System

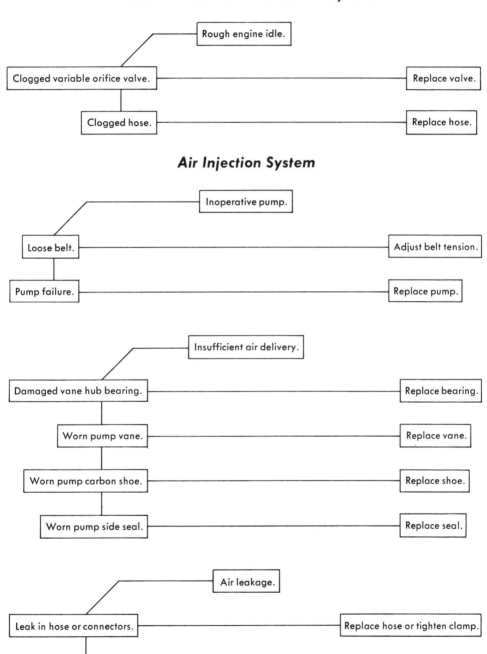

DATSUN

Air Injection System—Cont'd

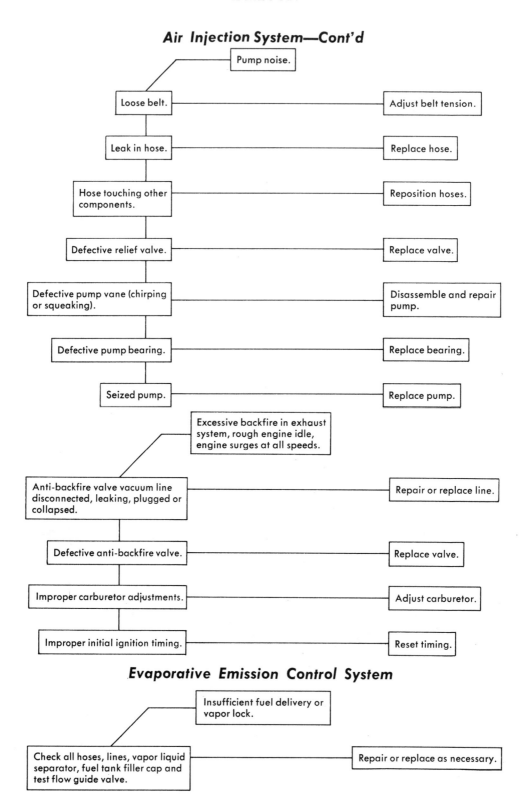

Pump noise.

Loose belt. ——————————— Adjust belt tension.

Leak in hose. ——————————— Replace hose.

Hose touching other components. ——————————— Reposition hoses.

Defective relief valve. ——————————— Replace valve.

Defective pump vane (chirping or squeaking). ——————————— Disassemble and repair pump.

Defective pump bearing. ——————————— Replace bearing.

Seized pump. ——————————— Replace pump.

Excessive backfire in exhaust system, rough engine idle, engine surges at all speeds.

Anti-backfire valve vacuum line disconnected, leaking, plugged or collapsed. ——————————— Repair or replace line.

Defective anti-backfire valve. ——————————— Replace valve.

Improper carburetor adjustments. ——————————— Adjust carburetor.

Improper initial ignition timing. ——————————— Reset timing.

Evaporative Emission Control System

Insufficient fuel delivery or vapor lock.

Check all hoses, lines, vapor liquid separator, fuel tank filler cap and test flow guide valve. ——————————— Repair or replace as necessary.